Biotechnology Monographs

Volume 2

Editors

S. Aiba · L. T. Fan · A. Fiechter · K. Schügerl

S. M. Stronach · T. Rudd · J. N. Lester

Anaerobic Digestion Processes in Industrial Wastewater Treatment

With 26 Figures

Springer-Verlag
Berlin Heidelberg New York Tokyo

Sandra M. Stronach, Thomasine Rudd and John N. Lester

Public Health Engineering Laboratory
Department of Civil Engineering
Imperial College, London, SW7 2BU, U.K.

ISBN 3-540-16557-6 Springer-Verlag Berlin Heidelberg New York Tokyo
ISBN 0-387-16557-6 Springer-Verlag New York Heidelberg Berlin Tokyo

Library of Congress Cataloging in Publication Data.
Stronach, Sandra Mary. Anaerobic digestion processes in industrial wastewater treatment.
(Biotechnology monographs; v. 2) Includes bibliographies and index. 1. Sewage sludge digestion.
2. Bacteria, Anaerobic. 3. Biomass energy. I. Rudd, Thomasine. II. Lester, John Norman. III. Title.
IV. Series.
TD769.S77 1986   628.3′51   86-6486
ISBN 0-387-16557-6 (U.S.)

Typesetting, printing and bookbinding: Brühlsche Universitätsdruckerei, Giessen
2152/3145-543210

Preface

There have been many significant microbiological, biochemical and technological advances made in the understanding and implementation of anaerobic digestion processes with respect to industrial and domestic wastewater treatment. Elucidation of the mechanisms of anaerobic degradation has permitted a greater control over the biological parameters of waste conversion and the technical advances achieved have reduced the time and land area requirements and increased the cost-effectiveness and efficiency of the various processes presently in use. By-product recovery in the form of utilisable methane gas has become increasingly feasible, while the development of new and superior anaerobic reactor designs with increased tolerance to toxic and shock loadings of concentrated effluents has established a potential for treating many extremely recalcitrant industrial wastestreams. The major anaerobic bioreactor systems and their applications and limitations are examined here, together with microbiological and biochemical aspects of anaerobic wastewater treatment processes.

London, June 1986   S. M. Stronach
                                                                 T. Rudd
                                                                J. N. Lester

Table of Contents

X

1 The Biochemistry of Anaerobic Digestion

Anaerobic digestion processes are widely used in the treatment of sewage sludge although the biochemical reactions comprising various stages in the anaerobic degradation of organic materials have not yet been fully elucidated. The overall anaerobic conversion of biodegradable organic solids to the end products CH_4 and CO_2 was initially believed to proceed in three stages which occurred simultaneously within the digester. These were:

1) the hydrolysis of insoluble biodegradable polymers;
2) the production of acid from smaller soluble organic molecules; and,
3) CH_4 generation.

The rate-limiting step in the digestion of soluble organic matter from the above scheme was considered to be the production of CH_4 from fatty acid degradation. The overall rate-limiting step for the complete anaerobic digestion of sludge was thus the hydrolysis by extracellular bacterial enzymes of insoluble polymeric molecules. Hobson [1] describes four interacting microbial reactions:

1) hydrolysis;
2) fermentation;
3) conversion of the products of acetogenic fermentation to CH_4; and,
4) conversion of the products of gaseous fermentation to CH_4.

Other intermediate, interrelating reactions occurring within the system during the digestion process include the formation and excretion of growth factors by one group of microorganisms that are necessary for the viability of other species, and the sequestration from the medium by one species of substances inhibitory to another. Gujer and Zehnder [2] proposed a six-step system in the anaerobic conversion of high molecular weight degradable organics to CH_4 and CO_2 (Fig. 1). These stages comprised:

1) hydrolysis of proteins, lipids and carbohydrates;
2) fermentation of sugars and amino acids;
3) anaerobic oxidation of long-chain fatty acids and alcohols;
4) anaerobic oxidation of intermediates such as the volatile fatty acids (with the exception of acetate);
5) conversion of acetate to CH_4; and,
6) conversion of H_2 to CH_4.

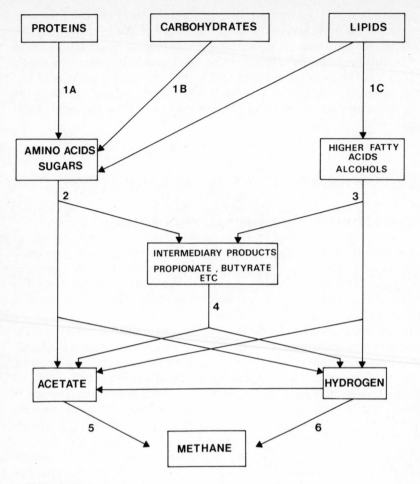

1. Hydrolysis
2. Fermentation
3. Anaerobic (β) Oxidation
4. Anaerobic Oxidation
5. Decarboxylation of Acetate $\quad CH_3COO^- + H_2O \longrightarrow CH_4 + HCO_3^-$
6. Hydrogen Oxidation $\quad\quad\quad CO_2 + 4H^+ \longrightarrow CH_4 + 2H_2O$

Fig. 1. Pathway of anaerobic biodegradation. (After Gujer and Zehnder [2] and others)

According to Eastman and Ferguson [3] lipids may not be hydrolysed in the first stage of an ongoing digestion process, but during the step of volatile acids fermentation.

Determination of the rate-limiting step in the anaerobic digestion of complex, insoluble biodegradable molecules may be problematical: the rates of degradation of such biopolymers depend not only upon their structures and substituent groups, but also upon the types of bacteria present in the digester and the efficiencies of these in substrate conversions under the conditions of influent flow rate, temperature and pH prevailing in the reactor. Kaspar and Wuhrmann [4] suggest that where soluble substrates are dominant, the rate-limiting reaction would tend to be the formation of CH_4 from acetate rather than from fatty acids under steady state conditions. These authors concluded that the large surplus capacity of H_2-consumption buffers the partial pressure of dissolved H_2, and stabilises the methanogenic ecosystems at values of pH_2 low enough to allow the rapid oxidation of fatty acids.

1.1 Kinetics of Substrate Utilisation and Bacterial Growth

1.1.1 COD Fluxes and Mean Carbon Oxidation State

In an anaerobic digester, the size of each group of bacteria will be in proportion to the amount of its particular substrate in the reactor system: the bacteria, or microbial groups, form the catalyst in each biologically-mediated reaction. The quantity of catalyst will not, however, limit the rate of the reaction; substrate or nutrient concentration will be rate-limiting.

The major gases formed in the degradation process are CH_4 and CO_2 and the fluxes of the diverse substrates or intermediary products are expressed as COD. If the main sink of electrons (or H_2) is carbon, as opposed to sulphur or nitrogen, then COD reduction results in the formation of CH_4 [2]. The volume of CO_2 involved is difficult to quantify as any estimation is complicated by the amount of bicarbonate formed from the gas. This conversion is dependent upon ammonia concentration and the proportion of gas dissolved in the digester fluid. If the substrate is completely mineralised, the mean oxidation state of the carbon in the organic material in the system will dictate the relative proportions of the components of the biogas. However, the nitrogen content of the degraded matter and the level of CO_2 saturation of the reactor medium will also influence the biogas composition to a limited extent.

Gujer and Zehnder [2] reported that the mean oxidation state of the digester sludge could be estimated from the following relationship:

$$\overline{OS} = 1.5 \frac{COD}{TOC} - 4, \qquad (1)$$

where $\overline{OS}$ = mean oxidation state of the carbon degraded
COD = amount of COD degraded (mass unit)
TOC = amount of organic carbon degraded (mass unit).
If the composition of the substrate is known, and total conversion of the substrate to biogas occurs, the theoretical CH_4 yield can be estimated from the equation of Symons and Buswell [5]:

$$C_nH_aO_b + \left[n - \frac{a}{4} - \frac{b}{2} \right] H_2O \rightarrow \left[\frac{n}{2} + \frac{a}{8} - \frac{b}{4} \right] CH_4 + \left[\frac{n}{2} - \frac{a}{8} + \frac{b}{4} \right] CO_2. \qquad (2)$$

A distinction should be made here between the terms wastewater and sludge in anaerobic digestion systems. In the former, the substrates for bacterial action are largely in the soluble state, whereas sludge contains much particulate organic material that requires hydrolysis by extracellular enzymes prior to bacterial assimilation.

1.1.2 Bacterial Growth and Biokinetics

1.1.2.1 Growth and Single Substrate Kinetics

The anaerobic conversion stages occurring in digesters have been analysed and expressed in terms of discrete reactions, as continuous cultures with one substrate that is limiting, but with an excess of nitrogen [1, 6]. Substrate utilisations can be described by simple Monod kinetics applicable to bacterial growth in pure culture chemostats. The growth rate of the biomass is assumed to be related to the biomass concentration:

$$\frac{dX_B}{dt} = \mu X_B,$$

(3)

where X_B = biomass concentration
μ = specific biomass growth rate.
The specific biomass growth rate is not a constant, but is related to the residual substrate concentration, S_2, in the system. The semi-empirical equation developed by Monod [7] predicts that:

$$\mu = \frac{\mu_{max} S_2}{K_s + S_2},$$

(4)

where μ_{max} = maximum specific biomass growth rate
K_s = half velocity coefficient.
K_s describes the substrate concentration which supports a specific growth rate that is half the maximum rate, i.e., it provides an indication of the ease of biomass assimilation of substrate (see Fig. 2).

Where one limiting dissolved substrate is involved, a basal model of degradation in digesters describes the residual substrate at different detention times in terms of the equation applied to completely mixed continuous cultures. At a dilution rate, D, at steady state, the residual limiting substrate concentration, S_2 (see Fig. 3) is given by:

$$S_2 = \frac{DK_s}{\mu_{max} - D}.$$

(5)

Under steady state digester conditions, the bacterial growth rate is equal to the dilution rate, i.e.:

$$\frac{dX_B}{dt} = 0$$

(6)

4

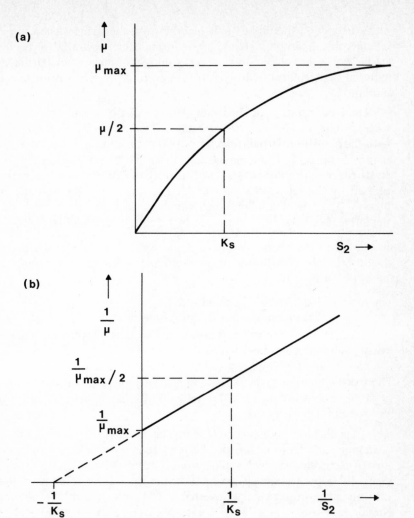

(a)

(b)

Fig. 2. Growth as a function of substrate concentration

(a) $\mu = f(S_2)$

(b) Lineweaver-Burk plot; Saturation Constant $K_s = S_2$ at $\mu = \mu_{max}/2$.
When the reciprocal of μ is plotted against the reciprocal of S_2, the straight line obtained has a slope of K_s/μ_{max}, an intercept of $1/\mu_{max}$ on the $1/\mu$ axis and an intercept of $-1/K_s$ on the $1/S_2$ axis. The double-reciprocal plot allows a much more accurate determination of μ_{max}, which can only be approximated as a limiting value at infinite substrate concentration from a simple plot of μ vs. S_2 [i.e. (a)]

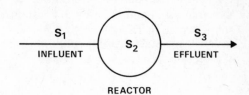

Fig. 3. Substrate concentrations in the digester

and:

$$\mu = D = \frac{\mu_{max}S_2}{(K_s + S_2)}. \tag{7}$$

The concentration of substrate available for bacterial assimilation governs, within limits, the growth rate of bacteria in the system; S_2, therefore, increases with the increase of growth rate imposed by the dilution rate, D. As the value of D approaches μ_{max}, the biomass begins to wash out of the vessel. The influent substrate concentration term, S_1, is not included in Eqs. (5) and (7), and the substrate utilised by the microorganisms $(S_1 - S_2)$ governs only the amount of bacterial growth. The rate of substrate removal or utilisation can, however, be related to the influent substrate concentration by an equation similar to (7):

$$k = \frac{r_{x,max}S_2}{K_s + S_2}, \tag{8}$$

where k   = substrate utilisation rate $[(S_1 - S_2)/T]$
   $r_{x,max}$ = maximum substrate utilisation rate.
   A constant relationship was also found to exist between the mass of bacteria produced and the amount of substrate utilised (7):

$$Y = \frac{\text{mass of bacteria formed}}{\text{mass of substrate removed}}, \tag{9}$$

were Y is the biomass yield coefficient. The biomass yield is of significance in the economic considerations of specific treatment processes, as it indicates the volume of excess sludge which will be produced and which will require treatment and disposal. A low yield coefficient is advantageous since it indicates that most of the influent carbon is being converted to CH_4 gas, and little resultant sludge accumulates. However, too low a biomass yield can also render the reactor sensitive to the effects of shock loading.

Substrate concentrations are initially expressed in grams or milligrammes per litre, as these units are generally used in descriptions of experimental digesters. The rate of substrate utilisation (r_x) can thus be expressed as kgCOD kg^{-1} VSS d^{-1} and the yield coefficient (Y) as kg VSS kg^{-1} COD. The dilution rate is the ratio of the flow rate of the pump, F, (1 h^{-1}) and the volume of the digester, V, (1), being the flow rate per unit volume, or F/V; in comparisons of theoretical and experimental results, the reciprocal of this parameter (1/D), the retention time, is conveniently used. The variation of key digester features with dilution rate is shown in Fig. 4 and the growth constants applicable to some anaerobic cultures are presented in Table 1.

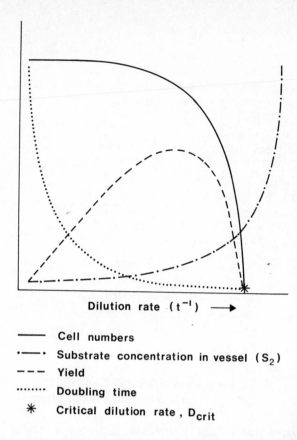

Dilution rate (t^{-1}) ⟶

——— Cell numbers

·——· Substrate concentration in vessel (S_2)

– – – Yield

········ Doubling time

✳ Critical dilution rate, D_{crit}

Cell numbers are maintained over a range of dilution rates; at the point where substrate influent rate exceeds cell growth rate (D_{crit}), organisms wash out exponentially and S_2 increases rapidly.

Fig. 4. Variation of key digester features with dilution rate

Table 1. Anaerobic growth constants. (After Henze and Harremoes [14])

Culture	μ_{max} 35°C d^{-1}	Y_{max} maximum yield coefficient kg VSS kg^{-1} COD	$r_{x,max}$, 35°C, kg COD kg^{-1} VSS·d^{-1}		K_s kg CODm^{-3}
			100% active VSS	50% active VSS	
Acetogenic bacteria	2.0	0.15	13	7	0.2
Methanogenic bacteria	0.4	0.03	13	7	0.05
Combined culture	0.4	0.18	2	1	–

1.1.2.2 Multisubstrate Systems

Very few, if any, digester influent streams are completely homogeneous. Where complex substrates are present, bacterial degradation will not occur at the same rate for every specific substrate. The cumulative effect, nevertheless, of all the conversions occurring in the system will generally produce a smooth curve of degree of degradation against retention time.

Hobson [1] observed that the degradation of solids at varying retention times in wastestreams of near constant composition often appeared graphically as biphasic curves [1, 8, 9] although application of Eq. (5) (above) gave a smooth curve for S_2 (residual limiting substrate at steady state) against retention time. The biphasic form was obtained, however, by treating the influent solids as comprising two bacterial substrates with dissimilar degradation kinetics.

In a reactor operating at a dilution rate of D, the concentration of residual solids remaining (X_2) could then be expressed as:

$$X_2 = X_{a,2} + X_{b,2} = \frac{DK_{s,a}}{(\mu_{max,a} - D)} + \frac{DK_{s,b}}{(\mu_{max,b} - D)}, \qquad (10)$$

where

X_2 = particulate residual substrate concentration

$X_{a,2}; X_{b,2}$ = particulate residual substrate concentration of microbial systems a and b

$K_{s,a}; K_{s,b}$ = half velocity coefficients of microbial systems a and b

$\mu_{max,a}; \mu_{max,b}$ = maximum specific biomass growth rates of systems a and b.

Two fractions of the biodegradable solids are thus being broken down by two populations of bacteria. In the model $(X_{a,1} - X_{a,2}) + (X_{b,1} - X_{b,2})$ are the particulates degraded, X_1 being the concentration of the degradable fraction of influent relevant to the system. Equation (10) describes the bacterial growth rates (μ), with actual values for μ_{max} and K_s. In practice, the particulate residual substrates concentration, X_2, should be much larger than the half-rate coefficient to maintain the steady state. The influent concentration X_1 cannot likewise be exceeded by X_2, and can limit the maximum value of μ attained to well below μ_{max}. The bacterial systems a and b have varying values of K_s, X_1, and μ_{max}; the dilution rate D, can therefore be greater than one or both μ_{max} values, or the maximum μ value. One bacterial system will be subject to washout as D increases, leaving its relevant substrate undigested; as D increases further, the second microbial population will wash out, resulting in the complete cessation of solids breakdown [1].

1.2 Kinetics and Biochemistry of Hydrolysis

As bacteria in general can only take up organic matter in soluble form, microbial assimilation of heterogeneous, particulate biopolymers requires breakdown or hydrolysis (see Fig. 1) as the first step. This process is mediated by extracellular enzymes, the reaction rates of which are influenced by pH, cell residence time and waste content of the digester. The pool of particulate material in the system must also be considered to include newly synthesised biomass [2]. Methane and CO_2 are produced principally as the result of the degradation of solubilised compounds, therefore the net decay rate of particulates and the accumulation rate of soluble compounds can be used to estimate the rate of gas production. Bacteria utilise the organic products of hydrolysis for net biomass synthesis, resulting in a net decay rate for the system that is less than the actual rate of hydrolysis. Hydrolysis rates could therefore be used to define circumstances under which operational instability might be expected, these rates allowing prediction of gas production and degree of stabilisation of organic matter.

In a digester system stabilising domestic sludge under steady state conditions (i.e. HRT > 12 days at 33 °C), the rates of production of soluble organic materials are balanced by similar removal rates, thus ensuring that there is little accumulation of these dissolved organics [10]. The soluble fraction of the total organic carbon (TOC) in the system is generally less than 10% of the whole (i.e. present as dissolved organic carbon or DOC). In an anaerobic digester with a mean HRT of 40 days at 33 °C, very short mean residence times of acetate (1 h) and propionate (0.7 h) were recorded [4]. The products of hydrolysis do not accumulate, and the distribution of the individual volatile acids has been demonstrated to be variable, illustrating the differences in the pathways of catabolism utilised by the digester bacteria: no change in the degree of particulate solubilisation occurred [3]. The anaerobic degradation of waste comprising easily-hydrolysable carbohydrates, such as those of sugar and starch-processing, may be subject to extremely rapid acidification [11]. If substrate conversion in such a system is governed by the rate of CH_4 formation, a sudden shock loading could lead to a concomitant increase in acid production, which could not be matched by a like increase in methanogenesis. The build-up of intermediate acids would cause a pH decrease, and further inhibition of CH_4 production would result.

The rate of biogas formation, with respect to the particulate organic matter remaining in the system that is available for conversion, has been shown to follow first order kinetics, with the net decay rate constant (k_p) being temperature dependent [1]. Other investigations have indicated that the solubilisation of particulate organic carbon in domestic sludge during the acidogenic phase of digestion is approximately a first order reaction at constant pH and temperature, in respect of the remaining biodegradable particulate substrate. There are limitations to the description of hydrolysis, however, by first order kinetics: Eastman and Ferguson [3] interpret this function as an empirical expression that reflects the cumulative effect of all the bacterial reactions occurring in the digester, the important implication of which is that different particulates are degraded at different rates. Hence non-degradable polymers such as waxes and lignins can delay the hydrolysis of associated particulate matter, by, for example, sterically hindering the activity of hydrolysing enzymes. Thus, although a first order function may be most suitable for complex substrates, the degradation of individual, homogeneous substrates may be best described by other hydrolysis functions. The breakdown of cellulose is an example.

9

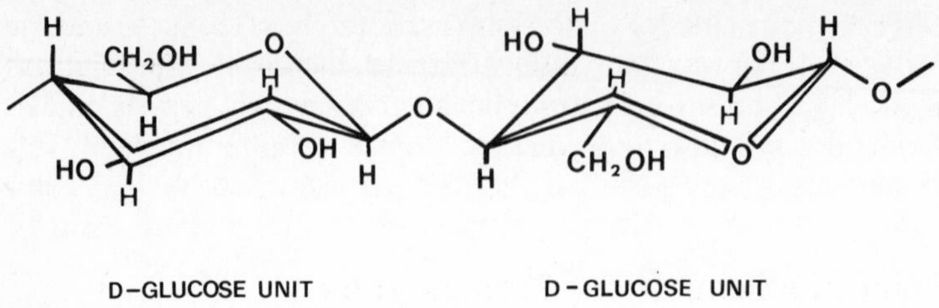

D—GLUCOSE UNIT   D–GLUCOSE UNIT

Fig. 5. Molecular structure of the cellobiose unit

Cellulose is a linear polymer of cellobiose units, connected by β-1,4-glucosidic linkages. Two anhydroglucose molecules make up the cellobiose unit (see Fig. 5). The degree of polymerisation (DP) of the D-glucose units in most celluloses is between fifty and several thousand. Cellodextrins with DPs of 1 to 6 are water soluble, but solubility decreases with increasing DP, and cellohexose (DP = 6) is only slightly soluble [12]. As cellulose chain length increases, chains in an aqueous environment associate. Hydrogen bonds and van der Waals forces become strong linking agents between cellulose molecules, causing adjacent chains to coalesce into tightly packed crystalline regions from which water molecules are excluded.

The hydrolysis of cellulose involves breakage of the chain by the addition of a water molecule to the β-1,4-linkage connecting two adjoining glucose molecules. The solubilisation by extracellular enzymes of native cellulose proceeds with a slight reduction in the degree of polymerisation and a conversion of substrate to soluble sugar. The rate of enzymic reaction appears to be a function of the cellulose surface area accessible to the cellulase enzyme and according to Cowling [13] is also influenced by:

1) the proportion of moisture in the fibre;
2) the degree of cellulose crystallinity;
3) the size and diffusional properties of molecules of enzyme relative to the size and surface characteristics of large capillaries and the spaces between cellulose molecules and microfibrils in amorphous regions;
4) unit cell dimensions of the cellulose;
5) the extent of polymerisation;
6) the conformational and steric properties of the anhydroglucose moieties;
7) the nature of the substances associated with the cellulose; and,
8) the concentration, distribution and nature of substituent groups.

Complete enzymic hydrolysis of cellulose requires the presence of 3 different enzyme components, known as cellobiase, C_1 and C_x, although several other names exist for these substituents [12]. In general, the C_1 component is believed to be an exoenzyme which splits off cellobiose units in an endwise manner, after the random activities of C_x enzymes have hydrolysed β-1,4-glucosidic linkages, i.e. the C_1 enzymes associate themselves with the free chain-ends which are generated by C_x activity and release cellobiose units, the major hydrolysis product. A β-glucosidase, cellobiase, catalyses the hydrolysis of cellobiose to glucose.

Most studies relate the kinetics of hydrolysis to the weight of cellulose and the initial cellulase activity in the reaction vessel. However, the kinetic formulations are complicated, are based on several assumptions and many rely heavily on Michaelis-Menten kinetics for their derivation [12].

1.3 Kinetics and Biochemistry of Fermentation and β-Oxidation

The breakdown products of the hydrolysis phase of biopolymer degradation form the substrates of the intermediary stages of the anaerobic digestion process (see Fig. 1). The acetogenic bacteria are the predominant microflora and acetate production is the principal result of their activity although higher volatile acids such as propionate, butyrate, iso-butyrate, valerate and iso-valerate may also be formed. The major routes of product formation by the acidogens are shown in Fig. 6. Their conversions are either fermentations or β-oxidation processes.

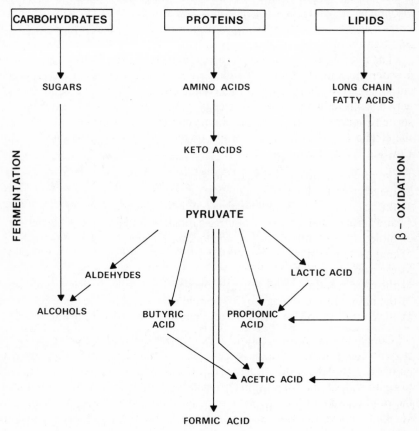

Fig. 6. The major routes of product formation by the acid-producing bacteria. (After Henze and Harremoes [4])

The concentration of fatty acids present under stable conditions in the digester tends to be low, of the order 0.1 to 0.3 kg CH_3COOH m^{-3}; high concentrations suggest near maximum loading conditions or operational instability [14]. The latter has been detected by estimations of the propionate concentration in the digester system [15]. The acidogenesis rate is normally much greater than the rate of methanogenesis; CH_4 formation, consequently, can be inhibited by the accumulation of volatile acids under increased loadings of soluble organic material.

Anaerobic microorganisms obtain the main part of their energy from oxidation-reduction reactions where electrons are transferred from one organic intermediate of sugar breakdown (the donor) to another organic intermediate (the acceptor) in the fermentation process. Sugars are in general the most common substrate for anaerobic fermentations, but some anaerobes can mediate the fermentation of fatty acids, amino acids, purines or pyrimidines [16]. Glycolysis and alcoholic fermentation are the predominant forms of glucose breakdown.

The β-oxidation process results in the successive removal of acetate (C_2) units from long chain fatty acids. The primary step in the reaction is the activation of a fatty acid by its transformation into the corresponding CoA thioester, an ATP-dependent process. β-oxidation is a cyclic process, with one mole of acetyl-CoA released at every complete turn of the cycle.

Fermentation may therefore be described as a microbiological reaction whereby organic compounds act as both electron donors and electron acceptors, whereas in anaerobic β-oxidation, the main sink for electrons is molecular H_2 (see Fig. 1). It has been established in several instances that removal of the electrons generated in glycolysis during acid-forming metabolism can occur via the mechanism of interspecies H_2 transfer, from acidogenic to methanogenic bacteria [17–19].

Hydrogen formed during the fermentation process (see Fig. 7) derives from the dehydrogenation of pyruvate (i.e. removal of two hydrogen atoms from pyruvate), a mechanism which is not inhibited by increased partial pressures of H_2 (ca. 101 kNm^{-2} H_2). During the anaerobic β-oxidation of fatty acids (i.e. oxidation at the β-carbon to yield a β-keto acid), the oxidation of reduced pyridine nucleotides and ferredoxin forms the source of H_2. Pyruvate dehydrogenation has a low redox potential (-0.68 V at pH 7.0), whereas the higher redox potential of NAD(P)H oxidation (-0.32 V at pH 7.0) results in inhibition of this process by elevated partial pressures of H_2. The assimilation of H_2 by the CO_2-reducing methanogenic bacteria stimulates the acidogens to utilise the ferredoxin-mediated pyruvate lyase reaction, whereby electrons of a sufficiently low redox potential to permit the reduction of protons to molecular H_2 are produced [20].

It has been proposed that coupled oxidation-reduction reactions between two amino acids – one acting as H_2 donor and the other as H_2 acceptor – may be a major occurrence in the anaerobic digestion of proteins or mixtures of amino acids. The conversion of some proteins to volatile acids was found to be unaffected in a co-system of amino acid-degrading bacteria and methanogens, when methanogenesis was inhibited by chloroform [21]. This is indicative of the degradation of the component amino acids of proteins in the anaerobic digester by coupled oxidation-reduction processes: protein breakdown may not be entirely dependent upon the activity of methanogenic bacteria as H_2 acceptors.

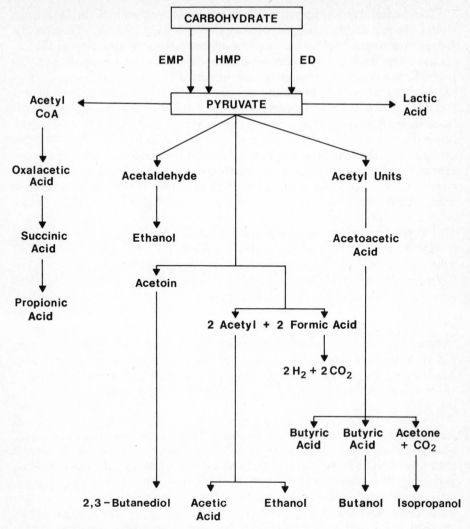

EMP = Embden-Meyerhof-Parnas pathway
HMP = Hexose Monophosphate pathway
ED = Entner-Doudoroff pathway

Fig. 7. End-products of the fermentation of carbohydrate

This coupled degradation of amino acids is generally referred to as the Stickland Reaction, and is believed to comprise a series of steps. The initial step requires the participation of the oxidised form of NAD^+ as a primary H_2 acceptor and an amino acid dehydrogenase system. The reoxidation of $NADH_2$ by the acceptor amino acid comprises the second step, together with an amino acid reductase system. ATP is generated during the formation of the carboxylic acid. A keto acid can replace the second amino acid in this reaction.

13

Clostridium propionicum utilises the Stickland pathway to form propionate from β-alanine: the amino group of β-alanine is initially transferred to pyruvate for the formation of α-alanine and malonate semialdehyde; an amino transferase catalyses this step. The malonate semialdehyde is used to form propionate and to aid the regeneration of β-alanine. The formation of acetate and CoA esters provides energy via ATP. Aminobutyrate, one C-atom longer than alanine, utilises the same system: *Cl. aminobutyricum* produces butyrate as the major end-product, the corresponding keto acid in aminobutyrate metabolism being 2-ketoglutarate [22].

The majority of the processes utilising single amino acids in anaerobic systems generate ammonia and are known as deaminations. These have several pathways, each dependent upon the enzymic complement of the organism and its environmental conditions:

1) Reductive deamination gives rise to the corresponding saturated fatty acid and ammonia; dehydrogenases catalyse the reaction and H_2 acts as H_2 donor:

$$R-CH_2-\underset{\underset{\displaystyle NH_2}{|}}{CH}-COOH + 2H^+ \rightarrow R-CH_2-CH_2-COOH + NH_3 \quad (11)$$

2) Desaturation deamination reactions generate the corresponding unsaturated fatty acid:

$$R-CH_2-\underset{\underset{\displaystyle NH_2}{|}}{CH}-COOH \rightarrow R-CH=CH-COOH + NH_3 \quad (12)$$

Arginine conversion, for example, is rapid. Deamination to citrulline by an arginine deaminase is the initial step; ornithine carbamoyl transferase catalyses a reversible transferase reaction to produce carbamoyl phosphate and ornithine. The energy-rich carbamoyl phosphate is ultimately degraded to ammonia, CO_2 and ATP. This formation of ATP is unusual in amino acid metabolism, but clostridia, mycoplasmas and streptococci can use the above process for arginine metabolism. The ornithine thus produced can be subsequently metabolised by *Cl. sticklandi* to yield acetate, propionate, valerate and butyrate, but this pathway has not yet been clearly established [22].

During the fermentation of glucose, acetate formation from acetyl phosphate via an acetate kinase reaction permits increased growth of the acid-producing bacteria, as substrate level phosphorylations result in additional ATP production:

$$ADP + \text{acetyl phosphate} \rightarrow ATP + \text{acetate.} \quad (13)$$

The operation of this mechanism by rumen bacteria yields acetate as the major fatty acid product, with lower concentrations of other fatty acids.

Growth characteristics were estimated for the fermentation of primary sludge: at $T = 35\,°C$, $pH = 5.2$, the maximum growth rate, $\mu_{max} = 2.7\,d^{-1}$; biomass yield

$Y = 0.48$ g cell g^{-1} COD utilised; decay coefficient $K_d = 0.43$ d^{-1} [2, 3]. At a retention time of 9 h in a completely stirred tank reactor (CSTR), the residual soluble COD of nitrogenous material was 7% of the degraded matter (340 mg l^{-1}) and that of carbohydrate was 12% (110 mg l^{-1}). It was concluded that fermentation processes are not close to saturation at retention times typical of anaerobic digesters (> 12 d at 35 °C) [2]. The operation of anaerobic reactors is not therefore limited by fermentative bacteria; the reaction is also largely independent of pH.

The anaerobic oxidation of fatty acids and alcohols (step 3 of Fig. 1) has been investigated by workers analysing the anaerobic treatment of methanolic wastes [23, 24]. Fermentation of methanol to acetate and thence to CH_4 and CO_2 via fatty acid intermediates (i.e., via stage 4 of Fig. 1) has also been recorded in an anaerobic upflow reactor, the intermediate formation being dependent upon digester pH [25].

During long chain fatty acid degradation, H_2 production inhibits the reaction [26]. Methanogenesis from H_2 (stage 6 of Fig. 1) is extremely dependent on pH, and hence H_2 accumulation at low pH inhibits anaerobic oxidation. Not all the intermediates present during sludge and wastewater treatment processes at steady state have been clearly identified and little is known of them as their concentrations are usually so low as to preclude extensive investigation. Around 70% of the CH_4 generated originates from acetate metabolism, but this figure includes the acetate formed by the degradation of higher fatty acids by the acetogenic bacteria. The role of higher VFAs as digestion intermediates has been examined. It was estimated that 15% of the CH_4 generated by anaerobic sludge digestion originated from propionate [4]; figures of 13% CH_4 produced from propionate and 8% from butyrate during mesophilic digestion of cattle waste have also been reported [27]. Butyric acid is probably degraded in a manner similar to that of the higher fatty acids caproate, caprylate, valerate, heptanoate and iso-heptanoate, using protons as the only electron acceptors [28]. Saturation of proton-reducing or H_2-transferring pathways by an increase in the glycolytic flux leading to an enhanced electron supply after, for example, the addition of high doses of glucose, leads to the utilisation of organic electron acceptors instead of protons; in consequence, organic electron-sink products such as proprionate, butyrate, ethanol or lactate are formed [20]. Evidence for the rapid accumulation of propionate after glucose addition to a stable digestion system with a subsequent low rate of removal has been presented [11]. If, however, acetate is the primary fatty acid intermediate present during steady state anaerobic digestion processes, bacteria or microbial consortia able to degrade other organic products [28, 29] may not exist in numbers sufficient to ensure the efficient removal of the organic overflow products of acidogenesis. Nonetheless, propionate is degraded to acetate, CO_2 and H_2, a reaction catalysed by at least one defined syntrophic culture [29]. This oxidative conversion of propionate may, however, proceed along a biochemical pathway different to that of higher fatty acid oxidation.

A complete anaerobic oxidation of acetate to CO_2 using elemental sulphur as an electron acceptor can be carried out, sulphur being reduced to sulphide [30]. One species of the strictly anaerobic sulphate reducing bacteria (*Desulfobacter postgatei*) is highly specialised in utilising only acetate as organic substrate [31].

The acetate is oxidized by the following reaction [20]:

$$CH_3COO^- + SO_4^{2-} \rightarrow 2\,HCO_3 + HS \quad [\Delta G^{0'} = -48\ kJ\ mol^{-1}]. \quad (14)$$

A thermophilic anaerobe which forms elemental sulphur from thiosulphate has also been reported [32], although the role of such an organism in anaerobic digestion has not yet been investigated.

1.4 Kinetics of Methanogenesis

As described previously, around 70% of the CH_4 generated in an anaerobic digester traces its origin from acetate. The reaction is a decarboxylation:

$$*CH_3COO^- + H_2O \rightarrow *CH_4 + HCO_3^-. \quad (15)$$

The acetate to CH_4 conversion (stage 5 in Fig. 1) has figured prominently in several investigations, not least because acetate is the dominant intermediary fermentation product in the decomposition of organic compounds under anaerobiosis in natural habitats [4]. The methanogenic breakdown of acetic acid has been kinetically analysed and four discrete phases distinguished [33]:

1) a rapid methanogenic rate increase upon acetate addition;
2) an exponential rate increase with time, reflecting the exponential growth of methanogenic bacteria;
3) an approximately constant rate of CH_4 production; and,
4) a rapid decline associated with acetate depletion.

The growth kinetics for acetate decarboxylating processes have been derived using enrichment cultures [34], the organism involved probably being *Methanosarcina* spp. from which similar kinetic parameters have since been obtained [35] (see Table 2).

The principal acetate cleaver in the majority of anaerobic digesters is likely to be *Methanothrix soehngenii*, however, as it has a high affinity (and thus a small

Table 2. Growth kinetics for methanogenesis from acetate

T (°C)	$\mu_{max}(d^{-1})$	$K_s(mg\,l^{-1}\,COD)$	Source	Organism
25	0.24	930	Lawrence and McCarty [35][a]	*Methanosarcina*
30	0.24	356	Lawrence and McCarty [35][a]	*Methanosarcina*
35	0.34	165	Lawrence and McCarty [35][a]	*Methanosarcina*
35	0.44[b]	250[b]	Smith and Mah [36][a]	*Methanosarcina*
37	0.43	369	Massey and Pohland [37]	*Methanosarcina* (?)
37	0.21[b]	241[b]	Wandrey and Aivasidis [38]	*M. barkerii*

[a] Figures derived by Gujer and Zehnder [2].
[b] Approximate values.

K) for acetate. This species has been reported in several reactors in the absence of *Methanosarcina*, and at HRTs greater than 15 days at 35 °C can successfully outcompete other bacterial systems [2]. At high substrate concentrations, however, comparison of the typical growth kinetics of *Methanosarcina* and *Methanothrix* spp. shows that the former will predominate.

In a full-scale anaerobic reactor assembly operating at a retention time of 40 days at 33 °C, the acetate decarboxylating system was estimated to be saturated to approximately 31 to 47%, figures in agreement with the growth kinetics of *M. soehngenii* [2]. This organism was, however, reported to be inactive at pH values below 6.8, with an optimum pH range of 7.4 to 7.3 [38]. Extreme temperature sensitivity was observed in the methanogen's short term activity, but an increase in growth rate by a factor of 2.05 per 10 °C in the range 15 to 35 °C was indicated in long term tests.

Methanogenesis from H_2 is also a pH-dependent conversion [39, 40]. The following reaction occurs:

$$CO_2 + 4H_2 \rightarrow CH_4 + 2H_2O. \tag{16}$$

All methanogens examined to date are capable of CH_4 formation by the oxidation of H_2 and the reduction of CO_2, and this metabolic feature unites the diverse species of methanogenic bacteria. In this reaction, a coenzyme present only in methanogens, known as coenzyme M (CoM), is involved in the transfer of methyl groups. CoM was shown by Taylor and Wolfe [41] to be a simple compound, 2-mercaptoethansulphonic acid (see Fig. 8) and was also proved to be the factor in rumen fluid necessary for the growth of *Methanobrevibacter ruminantium* [42].

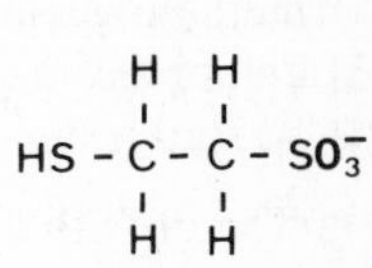

$$HS-\overset{\overset{\displaystyle H}{|}}{C}-\overset{\overset{\displaystyle H}{|}}{C}-SO_3^-$$

Fig. 8. 2-Mercaptoethanesulphonic acid   **Coenzyme M**

CoM is the smallest enzyme known, exceptional in its high sulphur content and acidity, and is required by methylcoenzyme M reductase, an enzyme universal in methanogens and active in the final steps of CO_2 reduction [43]. The exact mechanism of substrate conversion to CH_4 has not been elucidated, but methanogenic substrates may be bound to one or more carriers before being eventually reduced to CH_4 with regeneration of the carriers. The terminal reductive step of CH_4 formation in *Methanobacterium* strain MoH has been shown to occur as follows [41]:

$$CH_3\!-\!S\!-\!CoM \xrightarrow[\text{methyl reductase}]{H_2,\ Mg^{2+}ATP} CH_4 + H\!-\!S\!-\!CoM. \tag{17}$$

The ATP operates as the activator in the reaction [44] and the CoM as carrier.

Other coenzymes involved in methanogenesis have been found to be more structurally complex than CoM. Methanopterin may operate in the activation of

17

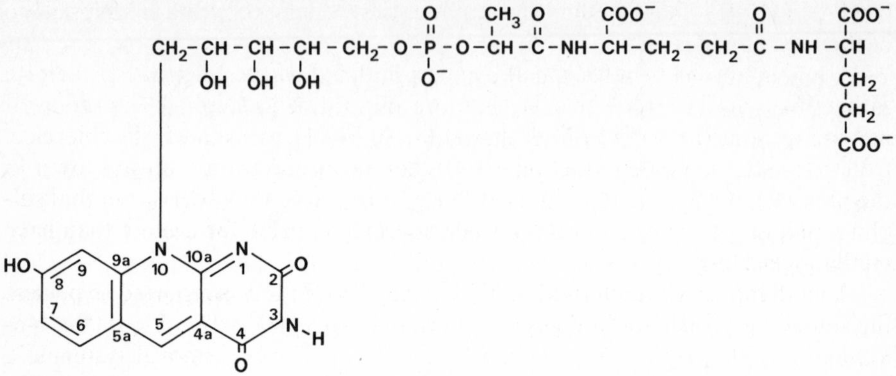

$$CH_2-CH-CH-CH-CH_2-O-\overset{\overset{O}{\|}}{\underset{\underset{O^-}{|}}{P}}-O-\overset{\overset{CH_3}{|}}{CH}-\overset{\overset{O}{\|}}{C}-NH-\overset{\overset{COO^-}{|}}{CH}-CH_2-CH_2-\overset{\overset{O}{\|}}{C}-NH-\overset{\overset{COO^-}{|}}{CH}$$

with OH OH OH on the first three CH groups, and the final CH bearing $CH_2-CH_2-COO^-$.

Fig. 9. Structure of F_{420}

CO$_2$, as carboxy-5,6,7,8-tetrahydromethanopterin (cTHMP) has been identified in methanogenic bacteria, suggesting that the carboxylation of methanopterin, with simultaneous reduction of the pterin molecule, may comprise the first step in methanogenesis from CO_2 [45, 46]. Methanopterin is a 2-amino-4-hydroxypterine with a complex side chain grouping [46, 48, 49].

Another entity, a fluorescent compound of low molecular weight known as factor$_{420}$ (F_{420}) is believed to be present in all methanogens, and functions as a primary electron carrier [49]. It was first described in relation to the H_2 metabolism of *Methanobacterium* strain MoH [50], and appears to be the flavin mononucleotide analogue 7,8-didemethyl-8-hydroxy-5-deazariboflavin-5'-phosphate with a phosphodiester-linked N-(N-L-lactyl-γ-L-glutamyl)-L-glutamic acid side chain attached [52] (see Fig. 9). The carrier acts as an electron transporter in the NADP-linked hydrogenase and formate dehydrogenase systems in methanogenic bacteria [52, 53]. The reduction of F_{420} by coenzyme-A-dependent pyruvate and α-ketoglutarate dehydrogenases in extracts of *Methanobacterium thermoautotrophicum* has been demonstrated [54], a reaction mediated by ferredoxin in most bacteria. When oxidised, F_{420} displays an adsorption peak at 420 nm whereas upon reduction both its adsorption at 420 nm and its fluorescence are lost [49].

Other chromophoric factors, designated F_{342} and F_{430} have been described in *M. thermoautotrophicum*, but their functions are unknown at present [55]. The latter is a nickel tetrapyrrole [56–58], the metal being coordinately bound within a uroporphinoid (Type III) ligand skeleton which additionally contains a carbocyclic ring and a chromophoric system.

Methane formation via H_2 and CO_2 has been reported to be inhibited by the presence of sulphate in sulphate-rich environments. The K_s values of a sulphate-reducing organism and an hydrogenotroph from sewage sludge were determined [17, 60, 61] and the rate of H_2 consumption by the sulphate-reducer (*Desulfovibrio vulgaris*) found to be five times that of the methanogen when the H_2 supply was rate-limiting [60]. Differences in substrate affinities thus account for the inhibition of CH_4 formation from H_2 and CO_2 in sulphate-rich environments with a low H_2 concentration. Methanogenesis and sulphate reduction are not mutually exclusive, however, and in the presence of excess H_2 have no effect on each other.

But when H_2 becomes rate-limiting, competition occurs, resulting in suppression of the methanogens.

A kinetic mechanism for the above was outlined, suggesting that the acetate rather than the H_2 route to CH_4 is more important in sulphate inhibition of methanogenesis (i.e., step 5 rather than step 6 of Fig. 1), as acetate is the chief electron donor for sulphate reduction [61]. Acetate concentration is always lower in the presence of sulphate than in its absence, supporting the observation that sulphate reducers have in general a lower apparent K_s value for acetate than have methanogens [64].

A methanogen characterised at pH 7.0 and $T = 33\,°C$ was reported to possess the following properties in digester supernatant: $\mu_{max} = 1.4\ d^{-1}$, $Y = 0.04$ g biomass g^{-1} COD [63] and $K_s = 0.6$ mg COD l^{-1} [2]. The H_2 removal system in a mature reactor (mean $HRT = 40$ d, $T = 33\,°C$) was found to function at about 1% of the potential rate intrinsic to the sludge ecosystem, results in accordance with reported growth kinetics [4].

According to Gujer and Zehnder [2], a thermodynamically stable anaerobic digestion system can only exist where propionate oxidation (step 4 of Fig. 1), acetate decarboxylation (step 5) and H_2 oxidation (step 6) are balanced. All three reactions must be exergonic, but the optimal range is extremely limited and governed by free propionate, acetate and H_2 concentrations. Mature digesters operate in the optimal range, commonly with concentrations of propionate and acetate between 10^{-4} and 10^{-3} molar, and partial pressures of H_2 no greater than $0.1\ kN\ m^{-2}$.

References

1. Hobson PN (1983) J Chem Technol Biotechnol 33B:1
2. Gujer W, Zehnder AJB (1983) Water Sci Technol 15:127
3. Eastman JA, Ferguson JF (1981) J Water Pollut Control Fed 53:352
4. Kaspar HF, Wuhrmann K (1978) Appl Environ Microbiol 36:1
5. Symons GE, Buswell AM (1933) J Amer Chem Soc 55:2028
6. Hobson PN, McDonald I (1980) J Chem Technol Biotechnol 30:405
7. Monod J (1949) Ann Rev Microbiol 3:371
8. Summers R, Bousfield S (1974) Agric Wastes 2:61
9. van Velsen AFM (1977) Neth J Agric Sci 25:151
10. Cohen A, Breure AM, van Andel JG, van Deursen A (1982) Water Res 16:449
11. Cohen A, van Deursen A, van Andel JG, Breure AM (1982) Antonie van Leeuwenhoek 48:337
12. Gilbert I, Tsao GT (1983) Interaction between solid substrate and cellulase enzymes in cellulose hydrolysis. In: Tsao GT (ed) Ann reports on fermentation processes, vol 6. Academic Press, New York London, p 323
13. Cowling EB (1975) Biotechnol Bioeng Symp 5:163
14. Henze M, Harremöes P (1983) Water Sci Technol 15:1
15. Kennedy KJ, van den Berg L (1982) Biotechnol Lett 4:137
16. Lehninger A (1975²) Biochemistry. Worth Publishers Inc, New York, p 417, 544
17. Hungate RE (1967) Arch Microbiol 59:158
18. Schiefinger CC, Linehan B, Wolin MJ (1975) Appl Microbiol 29:480
19. Winfrey MR, Nelson DR, Klevickis SC, Zeikus JG (1977) Appl Environ Microbiol 33:312
20. Thauer RK, Jungermann K, Decker K (1977) Bacteriol Rev 41:100
21. Nagase M, Matsuo T (1982) Biotechnol Bioeng 24:2227

22. Doelle HW (1979) Basic metabolic processes. In: Rehm H-J, Reed G (eds) Biotechnology: a comprehensive treatise in 8 volumes, 1: microbial fundamentals. Verlag Chemie, Weinheim-Deerfield Beach, Florida Basel, p 113
23. Lettinga G, van der Geest ATh, Hobma S, van der Laan J (1979) Water Res 13:725
24. Lettinga G, de Zeeuw W, Ouborg E (1981) Water Res 15:171
25. Adamse AD, Velzeboer CTM (1982) Antonie van Leeuwenhoek 48:305
26. Novak JT, Carlson DA (1970) J Water Pollut Control Fed 42:1932
27. Mackie RI, Bryant MP (1981) Appl Environ Microbiol 41:1363
28. McInerney MJ, Bryant MP, Pfennig N (1979) Arch Microbiol 122:129
29. Boon DR, Bryant MP (1980) Appl Environ Microbiol 40:626
30. Pfennig N, Bieble H (1976) Arch Microbiol 110:3
31. Brandis-Heep A, Gebhardt NA, Thauer RK, Widdell F, Pfennig N (1983) Arch Microbiol 136:222
32. Schink B, Zeikus JG (1983) J Gen Microbiol 129:1149
33. Powell GE, Hilton MG, Archer DB, Kirsop BH (1983) J Chem Technol Biotechnol 33B:209
34. Lawrence AE, McCarty PC (1969) J Water Pollut Control Fed Res Suppl 42:R1
35. Smith MR, Mah RA (1978) Appl Environ Microbiol 36:780
36. Massey MC, Pohland FG (1978) J Water Pollut Control Fed 50:2204
37. Wandrey C, Aivasidis A (1983) Ann NY Acad Sci 413:489
38. Huser BA, Wuhrmann K, Zehnder AJB (1982) Arch Microbiol 132:1
39. Huber H, Thomm M, König H, Thies G, Stetter KO (1982) Arch Microbiol 132:47
40. Wildgruber G, Thomm M, König H, Ober K, Ricciuto T, Stetter KO (1982) Arch Microbiol 132:31
41. Taylor CD, Wolfe RS (1974) J Biol Chem 249:4879
42. Taylor CD, McBride BC, Wolfe RS, Bryant MP (1974) J Bacteriol 120:974
43. Balch WE, Fox GE, Magrum LJ, Woese CR, Wolfe RS (1979) Microbiol Rev 43:260
44. Wolfe RS (1971) Microbial fermentation of methane. In: Rose AH, Wilkinson JF (eds) Advances in microbiological physiology, vol 6. Academic Press, New York London, p 107
45. Vogels GD, Keltjens JT, Hutten TJ, van der Drift C (1982) Coenzymes of methanogenic bacteria. In: Kandler O (ed) Archaebacteria. G. Fischer Verlag, Stuttgart New York, p 258
46. van Beelen P, Thiemessen HL, de Cock RM, Vogels GD (1983) FEMS Microbiol Lett 18:135
47. Keltjens JT, Caerteling CG, van Kooten AM, van Dijk HF, Vogels GD (1983) Arch Biochem Biophys 223:235
48. Keltjens JT, Huberts MJ, Laarhoven WH, Vogels GD (1983) Europ J Biochem 130:537
49. Zeikus JG (1977) Microbiol Rev 41:514
50. Cheeseman P, Toms-Wood A, Wolfe RS (1972) J Bacteriol 112:527
51. Eirich LD, Vogels GD, Wolfe RS (1978) Biochemistry 17:4583
52. Ferry JG, Wolfe RS (1977) Appl Environ Microbiol 34:371
53. Weimer PJ, Zeikus JG (1978) Arch Microbiol 119:49
54. Zeikus JG, Fuchs G, Kenealy W, Thauer RK (1977) J Bacteriol 132:604
55. Gunsalus RP, Wolfe RS (1978) FEMS Microbiol Lett 3:191
56. Diekert G, Jaenchen R, Thauer RK (1980) FEBS Lett 119:118
57. Jaenchen R, Diekert G, Thauer RK (1981) FEBS Lett 130:133
58. Gilles H, Thauer RK (1983) Europ J Biochem 135:109
59. Robinson JA, Strayer RF, Tiedje JM (1981) Appl Environ Microbiol 41:545
60. Kristjanssen JK, Schönheit P, Thauer RK (1982) Arch Microbiol 131:278
61. Schönheit P, Kristjanssen JK, Thauer RK (1982) Arch Microbiol 132:285
62. Zehnder AJB, Ingvorsen K, Marti T (1982) Microbiology of methane bacteria. In: Anaerobic digestion 1981. Elsevier Biomedical Press, Amsterdam New York Oxford, p 45
63. Zehnder AJB, Wuhrmann K (1977) Arch Microbiol 111:199

20

2 The Microbiology of Anaerobic Digestion

The combined and coordinated metabolic activity of an anaerobic reactor population is required for the complete degradation of complex organic matter to CO_2 and CH_4. The intermediates necessary for certain microorganisms are produced as a consequence of the action of others and therefore consortia of bacteria are frequently involved in these conversions. Despite several analyses of the major non-methanogenic bacteria present in anaerobic digesters, detailed investigations into the generic and specific nature of the hydrolytic and fermentative populations have not generally been reported. The predominant organisms in some waste-treatment systems may not, moreover, participate actively in the process but may merely be components of the wastestream itself; coliforms have been implicated here [1].

2.1 Nutrient Balance in Anaerobic Digesters

Efficient digestion processes require that the medium in which the microorganisms grow and multiply contains energy sources, sources of carbon and nitrogen for the biosynthesis of new cells, and trace elements, sulphur and other ions necessary for bacterial metabolism; cell carbon must come from carbohydrate, CO_2 and VFAs (including acetate for methanogens) [2].

The wastestream under treatment will generally contain the necessary nutrients unless the waste is highly specific in nature, but the main requirements, the nitrogen and energy constituents, are rarely present in the requisite proportions for optimal utilisation. The major source of nitrogen is ammonia from wastestreams, or from the hydrolysis of protein and the deamination of constituent amino acids, or from the hydrolysis of non-protein nitrogenous compounds such as urea, to ammonia. Simple sugars do not generally form part of typical anaerobic reactor feedstocks, unless the waste is derived from a process in which these are utilised or produced; carbohydrate is frequently found as cellulose, hemicellulose and other vegetable polymers, which may be fibrous, heavily-lignified and not readily degraded by the anaerobic hydrolysing bacteria [3]. The phosphorus requirement is considered to be low in the majority of systems [4], although the phosphate content of some factory effluents may limit bacterial growth: phosphate is necessary for nucleic acid synthesis and is an integral component of many other cellular constituents.

Several wastes, such as those of animal origin, tend to contain high proportions of nitrogen, whereas starch and sugar effluents may have proportionately little nitrogen. The C:N ratio can be adjusted by the addition of a co-substrate: the elevated ammonia content of some animal wastes, which remains unaltered during the digestion process, can be counteracted by supplementation with a starchy potato waste [3]. The optimisation of C:N ratios in wastestreams of animal origin has been investigated by Hills [5]. Values of the effective C:N ratios in feedstocks cannot be provided by analyses of cellulose and similar components as the proportions of carbon and nitrogen present in a form assimilable by the digester population cannot be analytically determined [3]. The starch component of potato waste for example, is amenable to the action of the amylase enzymes of some hydrolysing bacteria, whereas cellulose and hemicellulose are less easily degradable.

The proportions of nitrogen and phosphorus in the volatile SS formed as a result of the degradation cycle have been estimated as approximately 10.5% and 1.5% respectively [6]. The levels of essential nutrients in the influent can be evaluated if the biomass yield coefficient is known; the COD:N ratio is frequently utilised to describe nutrient requirement. For those anaerobic processes at high loadings (0.8–1.2 kg COD kg^{-1} VSS d^{-1}), a COD:N ratio of around 400:7 has been estimated [4, 7] whereas at lower loadings (<0.5 kg COD kg^{-1} VSS d^{-1}), values of 1000:7 or more may be necessary. The N:P ratio has been reported to be approximately 7 [6].

The sulphur requirement of anaerobic reactor systems is similar to that of phosphorus [7]. Other trace elements defined as necessary to the anaerobic digestion process include the free ionic species of iron (Fe^{2+}), nickel, magnesium, calcium, barium and cobalt [6, 8–10].

The nature of the wastestream under treatment has also been demonstrated to affect biogas composition, although the proportion of CH_4 present in the off-gas is frequently greater than estimated due to the dissolution of significant quantities of CO_2 in the digesting liquor. The contribution of protein to gas production appeared negligible in one investigation as the CH_4 produced from the anaerobic degradation of both protein-containing (14.5 g l^{-1}) and deproteinized potato waste was reported to be 0.3 m^3 kg^{-1} COD removed [11]. Complete carbohydrate breakdown, however, gives a 50:50 mixture of CH_4 and CO_2 whilst degradation of lipid gives a greater proportion of CH_4 [3].

2.2 Origin and Nature of Digester Bacteria

The treatment of waste products, in contrast to other industrial microbiological procedures, does not require conditions of complete sterility or pure microbial cultures. The wastestream itself may comprise the initial source of bacterial inoculum for the digestion process and may also act as a continuous source of fresh biomass whose introduction into the system could aid the stabilisation of the reactor flora: strain degeneration may reduce reactor efficiency where such reinoculation does not occur [3]. As noted above, some component bacteria in the system may not be requisite for the digestion process, but may nevertheless produce growth factors necessary to the degradative organisms or sequester toxic by-products of anaerobic metabolism. The heterogeneous reactor flora provide not only the microbial substrates for subsequent phases of the breakdown process but also contribute to anaerobiosis: a proportion of the population comprises facultative anaerobic microorganisms which, under conditions where air may have been inadvertantly introduced, can reduce any oxygen present and stabilise the Eh of the system [3]: the particularly fastidious methanogenic bacteria require an environmental Eh of around -300 mV to metabolise and proliferate.

A greater proportion of obligate, in comparison to facultative, anaerobes has been frequently isolated from anaerobic sludge digestion systems: obligate organisms have been reported to be present at 10–100 times the numbers of facultative bacteria, the results being attributed to the low Eh value of the environment [12]. In digesters treating animal wastes however, facultative anaerobes comprised approximately half of the isolates obtained; these indicated the cellulosic nature of the influent and the hydrolytic functions of the digester flora [13]. The component microorganisms of a particular anaerobic regime are influenced by the reactor feedstock: the facultative population of a sewage digester was found to consist mainly of *Escherichia coli*, a dominant coliform bacterium in the human digestive system, but upon acclimation to pig-waste as feedstock, the predominant facultative anaerobes subsequently isolated proved to be the streptococci of the porcine gut [13].

The majority of investigations into the microbiology of anaerobic digestions are based on mesophilic systems and although the bacterial types may differ in thermophilic operations, similar reactions probably occur. Of the total population of a mesophilic digester, 9% were found to be thermophilic (55 °C) and 1% proved to be obligate thermophiles (60 °C) [14].

Throughout the start-up phase of an anaerobic digestion process, the heterogeneous groups of microorganisms evolve in succession until the biocenosis reaches stabilisation and the various groups attain their final proportions. The composition of the influent stream will dictate to a marked degree the micropopulations which develop within the system. The progression of a digestion system initiated from water and pig waste was followed in a conventional anaerobic reactor [13]. During the first week of the process the only bacteria, of those groups possessing the hydrolytic and methanogenic capabilities to degrade the influent substrate, present in detectable numbers were those with amylolytic (starch-degrading) ability. The hydrolysis of the more complex components of the waste by extracellular enzymic activity into more amenable substances assimilable by a greater number of species forms the initial stage of the degradation of large polymeric molecules. The component substrates of the waste utilised in the above investigation were limited in range and comprised residual plant matter, undigested starch, bacterial and animal cells, secretions, nitrogenous compounds of urine and salts from animal feed; the large numbers of indigenous facultative microflora present resulted in the dominance of facultative over obligate anaerobes in the initial digestion stages (cf. [12]). Those bacteria hydrolysing starch had attained stable population numbers of approximately 4×10^4 cells per ml of digester

Table 3. Bacterial counts (per ml) during establishment of digestion by the slow addition of piggery waste to a 15l digester originally containing water. (After Hobson and Shaw [13])

Week	Obligate anaerobes						Facultative anaerobes
	Hydrolysers		Methanogens utilising				
	Cellulolytic	Amylolytic	Formate	Acetate	Propionate	Butyrate	
0	–	–	–	–	–	–	
1	–	$>4 \times 10^3$	–	–	–	–	2.2×10^7
2	–	$>4 \times 10^3$	–	–	–	–	2×10^6
3	–	$>4 \times 10^4$	–	–	–	$>2 \times 10^3$	9×10^7
4	–	$>4 \times 10^4$	$>2 \times 10^4$	–	–	$>2 \times 10^4$	3.2×10^7
5	$>4 \times 10^3$	$>2 \times 10^4$	$>2 \times 10^4$	–	–	$>2 \times 10^4$	3.2×10^7
6	–	–	$>2 \times 10^4$	–	–	$>2 \times 10^4$	5.1×10^6
7	–	–	$>2 \times 10^4$	–	–	$>2 \times 10^3$	7.1×10^6

Key: – no measurable count

fluid by the third week of system operation. The alteration of the influent to a waste containing an increased proportion of cellulose was compensated for by an increase in the numbers of cellulolytic microorganisms present (see Table 3). Methanogenic bacteria were not detected in the pig waste used to initiate the digestion system, but butyrate-assimilating methanogens were reported after the third week of the regime and formate-utilisers were present by week four. The successive changes in numbers and types of bacterial subgroups reflect the pattern of degradation steps in the breakdown of complex substrates, as well as the quantities of different substrates becoming available via the action of various microbial species. The consistent isolation of facultatively anaerobic bacteria (see Table 3) was probably indicative of the role of these organisms in the fermentation of the simpler sugars produced by the degradation of the complex carbohydrate substrate. The maintenance of the anaerobic digester at an Eh value conducive to the growth and metabolic activity of obligate anaerobes such as the methanogens and the hydrolysing bacteria is also considered to be a function of these subpopulations [13, 15].

Lipolytic bacteria i.e. those possessing the capacity to degrade lipids, were not tested for in the system described. However, some of the isolates obtained could ferment glycerol, a by-product of the hydrolysis of lipid, while others could utilise ammonia as their sole source of nitrogen, despite the reported maintenance of the ammonia component of the waste throughout the period of digestion. Under those conditions prevalent in an anaerobic reactor system, preferential utilisation of amino acids over ammonia may occur or an ammonium cycle may exist whereby the formation of ammonia is in equilibrium with its utilisation. The isolation of lactate and succinate-producing microorganisms was reported although these fatty acids were not detected in the digester [15], possibly due to the transient nature of their existence under the prevalent conditions. These acids are further metabolised under anaerobiosis to acetic and propionic acids by fermentative bacteria. Prior to the initiation of CH_4 production (which began in the third week) some higher fatty acids were formed, the predominant one being acetate [13, 15], the major fatty acid product of anaerobic digestion of complex substrates [16] and an important substrate of many methanogenic bacteria [17, 18].

The formation of formate, or CO_2 and H_2 occurs concomitantly with acetate production, the formate being a substrate of the methanogen *Methanobacterium formicicum* which was isolated in the digestion initiation experiments discussed above [13]. In a progressive anaerobic reactor system, the onset of methanogenesis tends to be accompanied by a decrease in the concentration of acetate.

Throughout the period of the establishment of a digestion process, comparative analyses of the numbers and activities of various groups of anaerobic microorganisms degrading different complex waste materials should not only provide an indication of the various types and quantities of substrate available for conversion during the period of digestion, but should reflect the degree of recalcitrance of the particular substrate under treatment. The persistence of one particular bacterial group, or one specific organic by-product may not only provide insight into the rate-limiting functions of various degradable materials but may also aid the estimation of digester efficiency.

2.3 The Hydrolysing Bacteria

Hydrolysis of macromolecules such as lipids, proteins and carbohydrates under anaerobic reactor conditions is primarily an extracellular enzymic reaction, and many microorganisms produce extracellular enzymes, chiefly hydrolases. The function of the hydrolytic enzymes, such as lipases, proteases and cellulases, is the degradation of complex molecules into units which can be assimilated by the microbial cell. In an anaerobic digestion process where organic polymers form a substantial portion of the wastestream to be treated, the hydrolysing bacteria and their enzymes are of paramount importance because their activity produces the simpler substrates for the succeeding steps in the degradation sequence (see Fig. 10).

Lipid hydrolysis under anaerobic digester conditions has not been extensively documented. The extracellular enzymes responsible for the conversion of the glycerol esters of lipids to long chain fatty acids are the lipases, important in the treatment of dairy wastes, for example. In mixed cultures of bacteria, some lipases may be inhibited by proteases. Most microbial lipases attack the 1 and 3 positions on the glyceride, but some can attack all three positions to give free fatty acids and glycerol [19]. Estimations of 10^4–10^5 lipolytic bacteria per ml of digester fluid have been recorded [3] and enrichment techniques have also resulted in the isolation of lipolytic microorganisms [20]. The clostridia and the micrococci appear to be responsible for the majority of the extracellular lipases in anaerobic digesters.

Proteolytic bacteria fulfil an important role in the stabilisation of raw sewage sludge. The most numerous proteolytic organisms isolated from digesters have

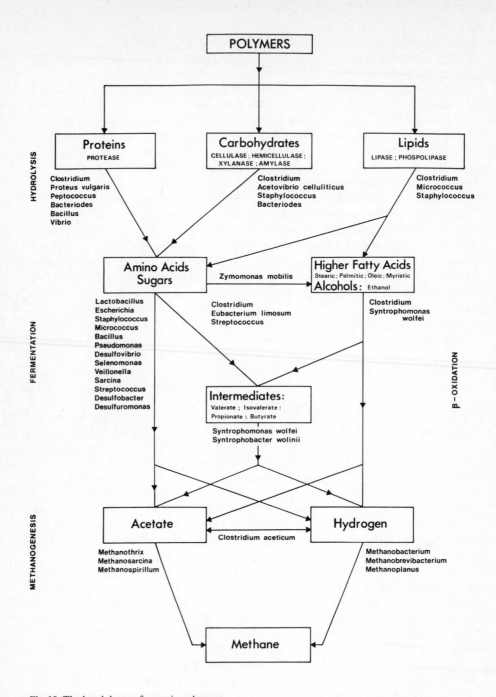

Fig. 10. The breakdown of organic polymers

26

been clostridia and anaerobic cocci [13, 21]; 10^4–10^6 proteolytic bacteria per ml of digester liquor have been reported. These included *Clostridium bifermentans, Cl. butyricum, Cl. perfringens, Cl. mangenotii, Cl. litusburense, Peptococcus anaerobius,* and *Staphylococcus aureus.* Other bacteria recorded belong to the genera *Sarcina, Bacteriodes,* and *Propionibacterium* [1].

The proteolytic enzymes of the hydrolytic bacteria are stable in the pH range 5.0–11.0 and have various optima [22]. Their actions range from the highly specific, such as the protease elaborated by the anaerobe *Cl. histolyticum* (a carboxypeptidase), which attacks the basic amino acid residues at the carboxyl side of the scission point, to the broad range, for example, the proteolytic enzymes of the streptococci. Proteases active in the neutral pH range (7.0–8.0) tend to be zinc-containing metallo-proteins that are subject to inhibition by metal-chelating agents such as ethylenediaminetetraacetic acid (EDTA). These reagents bind initially to the zinc atom without causing its detachment from the protease. Alkaline proteolytic enzymes such as those of many *Bacillus* species are not generally deactivated by metal-chelating agents, and share several properties: a serine residue at the active site, esterolytic activity, alkaline pH optima and sensitivity to organophosphorus reagents [22].

Most of the proteolytic isolates of one investigation were additionally able to hydrolyse carbohydrate, with gas and acid as the products [21]. This indicates the versatility of some of the hydrolysing bacteria found in the anaerobic digester system, where the ability to break down and catabolise more than one substrate must confer a selective advantage.

The microbiology of the breakdown of fibrous polysaccharide is subject to some controversy, as the degradation pathways are not fully understood. Eleven types of mesophilic cellulolytic bacteria were found in an anaerobic digester fed on pig-waste; all but one of these were Gram-negative [13]. All were likewise strict anaerobes and counts of the order of 4×10^5 organisms per ml of digester fluid were recorded. The thermophilic cellulolytic bacterium *Cl. thermocellum* is capable of the conversion of cellulose to ethanol, acetic acid and fermentable sugars [23].

A number of bacteria capable of growth on both native and modified celluloses do not elaborate cellulases in isolation in culture media. Some microorganisms produce only endo-glucanases or β-glucosidase and can therefore hydrolyse only soluble cellulase derivatives such as carboxymethylcellulose. Only those strains producing exoglucanase activity, i.e., C_1 (see Chap. 1) are capable of hydrolysing native cellulose. *Cl. thermocellum* was found to elaborate both C_1 and C_x cellulase activity, both components appearing to be totally extracellular [24]. Cellulase activity was not observed to occur in cellulose media which had been supplemented with high levels of cellobiose or glucose however, and it was deduced that this phenomenon represented the repression of cellulase synthesis rather than the inhibition of cellulase activity.

In addition to the cellulolytic bacteria in the anaerobic digester, microorganisms degrading hemicelluloses are also important in the initial phase of digestion. Two types of hemicellulolytic organism were isolated in a pig-waste anaerobic digestion [13], the dominant form being identified with the rumen bacterium *Bacteroides ruminicola,* the other a Gram-negative rod. Unlike the main group of

Table 4. Some important starch-degrading enzymes. (After Fogarty and Kelly [22])

Trivial name	Systematic name	Typical amylolytic microorganisms
α-Amylase	α-1,4-Glucan 4-glucanhydrolase	*Bacillus subtilis* *B. stearothermophilus* *B. licheniformis*
β-Amylase	α-1,4-Glucan maltohydrolase	*B. subtilis* *B. cereus*
Amyloglucosidase	α-1,4-Glucan glucohydrolase	Fungi
Debranching enzyme	α-1,4-Glucan 6-glucohydrolase	*B. cereus* *Escherichia coli*

hemicellulose-fermenting bacteria, neither of these isolates could grow with ammonia as the sole source of nitrogen.

The technique of culture enrichment has been used in an attempt to clarify some of the microbiological and biochemical aspects of cellulose degradation [25–27]. A more fundamental approach to the investigation of the process has been taken by co-culturing pure cultures of the microorganisms involved, i.e., methanogens with fermenting bacteria, or hydrolysers [28, 29].

The amylases are enzymes capable of the degradation of starch, glycogen and related polysaccharides by hydrolytic cleavage of α-1,4- and/or α-1,6-glucosidic linkages. Amylolytic bacteria reported in anaerobic digester processes include *Cl. butyricum*, *Bacteroides* spp., *Lactobacillus* spp. [13], *Bacillus subtilis*, *B. cereus*, and *B. licheniformis* [1]. Numbers of 4×10^4 per ml of medium have been recorded [13].

The amylases of most importance in biotechnological processes constitute four groups (see Table 4). The α-amylases mediate endocleavage of substrates; the β-amylases hydrolyse alternate bonds from the non-reducing end of the substrate; the amyloglucosidases hydrolyse successive bonds from the non-reducing end of the substrate, while the debranching enzymes cleave the α-1,6-glucosidic linkages in amylopectin and glycogen. All α-amylases are considered to be calcium metallo-enzymes. β-amylase has only recently been demonstrated in microorganisms and the enzyme system is subject to catabolic repression [22].

Other extracellular enzymes which may participate in the initial hydrolysis step of anaerobic digestion are the pectinolytic enzymes such as those elaborated by some *Bacillus* and *Clostridium* species, and the dextranases of, for example, *Bacillus* species.

2.3.1 End-Product Inhibition During Hydrolysis

Although distinct differences exist between the various hydrolysing bacteria capable of activity in anaerobic reactor systems, their synthesis of a number of extracellular enzymes seems to be governed, at least in part, by the concentrations of substrate and product in the digester liquor. It has been suggested that high concentrations of the products of hydrolysis inhibit the hydrolytic enzymes of the relevant bacteria [30]. Catabolite repression of enzyme synthesis is not unknown; synthesis of the α-amylase of *B. licheniformis* and *B. subtilis* is repressed by glucose, and by fructose in the case of *B. stearothermophilus*. The production of proteolytic enzymes is repressed by amino acids in many microorganisms but the proteases may also be repressed by glucose [22]. However, some extracellular enzymes are also inducible. End-product inhibition of extracellular proteases by amino acids is known and the specific repressive amino acids identified in a number of cases [22].

2.4 Intermediate Metabolism

2.4.1 The Fermenting Bacteria

The second stage in Gujer and Zehnder's [31] general outline of the digestion process is the fermentation of amino acids and sugars, giving rise to the intermediary products and acetate or H_2 (see Fig. 10). Acetate is the most important compound quantitatively produced in the fermentation of organic substrates by bacterial populations, with propionate production of secondary consequence [32, 33]. Only a limited number of microorganisms have been reported to mediate the further degradation of propionate under anaerobic conditions [34]. The breakdown of amino acids under conditions of anaerobiosis has been examined in terms of the interactions between methanogenic and amino acid degrading bacteria [35, 36]: certain amino acids were reportedly degraded oxidatively (cf. Fig. 10) by dehydrogenation, with the associated methanogenic bacteria acting as H_2-acceptors. The inhibition of methanogenesis by chloroform [37, 38] also inhibited the degradation of the specified amino acids and/or caused variations in the composition of the VFAs produced from them. Further experimental insight was gained into the interactions of these OHPA bacteria and methanogens by the addition of glycine to the system; the latter reduced the inhibitory effect of chloroform, probably operating as the H_2-acceptor in place of the methanogenic bacteria [36].

Acidic end-products and ammonia are formed from the amino or amide groups of amino acids. The catabolism of these organic compounds is mediated by a large number of both obligatory and facultatively anaerobic microorganisms and the process utilises single amino acids, pairs of amino acids or a single amino acid in conjunction with a non-nitrogenous compound (see Chap. 1). The conversion of single amino acids by reductive or desaturation deaminations is carried out under anaerobic conditions by clostridia, mycoplasmas and streptococci. The amino acid arginine is metabolised to ammonia, CO_2 and ATP (see Chap. 1), ornithine to acetate, propionate, valerate and butyrate, and lysine mainly to acetate and butyrate. The metabolism of the latter amino acid is singular in that a shift of the amino group on the molecule occurs prior to the deamination reaction. A number of clostridia utilising protein hydrolysates or mixtures of amino acids seem to acquire the greater part of their energy by a coupled oxidation-reduction reaction between appropriate amino acids. *Cl. propionicum* and *Cl. butyricum* use the Stickland Reaction in the coupled interaction of a single amino acid and a keto acid [39].

Temperature, pH and the composition and nutrient quality of the influent feed are crucial to end-product formation. Anaerobic growth in heterotrophic microorganisms presents a particular difficulty in that the total ATP requirement of these bacteria for biosynthesis can only be accommodated by the degradation of an organic energy source in relatively large quantities. There are a number of specific control mechanisms in strict and facultative anaerobes which govern the flow of electrons; one of these is the enzymic removal of excess electrons in the form of molecular H_2 via the action of an hydrogenase. This potential for production of H_2 in molecular form is prevalent, but not ubiquitous; where it does not occur (e.g., in iron-deficient media, or by lack of the requisite enzyme), the organisms are restricted to more conventional fermentation reactions.

The fermentation of substrates in anaerobic digestion systems to yield acetone, butanol, butyric acid and iso-propanol is mediated predominantly by bacteria of the genera *Clostridium* and *Butyribacterium*. The formation of butyrate as the main end-product of fermentation is generally a conversion process confined to obligate anaerobes. Major end-product formation in the clostridia has provided a means of classification: *Cl. butyricum* therefore produces butyrate, *Cl. acetobutylicum* mainly acetone and butanol, and *Cl. butylicum* predominately butanol, in addition to H_2 and CO_2. The conversion of sugars to pyruvate via the Embden-Meyerhof-Parnas (EMP) pathway initiates the butyric acid fermentation.

Pyruvate breakdown is characteristic of the clostridia and is often referred to as the "clostridial type". Pyruvate is decarboxylated to acetyl-CoA, CO_2 and H_2 via a pyruvate-ferredoxin oxidoreductase; the lower ferredoxin redox potential preserves anaerobic conditions and prevents additional transportation of electrons [16]. The acetyl CoA is maintained in equilibrium with acetyl phosphate by the enzyme phosphoacetyl transferase. The acetokinase enzyme then converts the acetyl phosphate to acetate, with the production of one mole of ATP [39]. One mole of glucose produces, therefore, 2 moles of acetate at the most, with 2 moles of CO_2 and H_2 evolved during pyruvate degradation. Two clostridial strains are known to have the ability to convert one mole of glucose into 3 moles of acetate: *Cl. thermoaceticum* and *Cl. formicoaceticum* do not release CO_2 and H_2; the third mole of acetate is formed by the reduction of CO_2 by the H_2 from the pyruvate reaction.

In the anaerobic digestion system, acetic acid as the sole end product is not conducive to stability as it causes a reduction in the pH value, rendering the reoxidation of $NADH + H^+$ more difficult. Members of the genus *Clostridium* also utilise pyruvate to form the weaker acid butyrate. Several saccharolytic clostridia normally producing butyrate from carbohydrates can alter their metabolism in such a way that acetone and butanol are produced if the formation of acid lowers the environmental pH to about 4.0. This conversion is unlikely under normal anaerobic reactor conditions unless the wastestream is highly acidic, but represents a possible stabilising influence. *Cl. acetobutylicum* possesses this capability and *Cl. butylicum* can reduce acetone further, to iso-propanol.

30

The presence of propionic acid and succinic acid are frequently reported under anaerobic digester conditions; these are products of carbohydrate or lactate fermentations, and the anaerobic microorganisms capable of these reactions include *Megasphaera* (formerly *Peptostreptococcus*), *Cl. propionicum*, *Propionibacterium pentosaceum*, and *Prop. shermanii*. It is generally observed that those bacteria which can metabolise glucose can also utilise lactate, although lactate-utilisers are not necessarily also glucose-utilisers. The propionibacteria prefer glucose as their source of carbon and energy, but *Megasphaera* spp. cannot metabolise this sugar, and utilise lactate.

In an animal waste digester, lactate-fermenting bacteria were reported in numbers of 3×10^4 per ml of digester fluid [3]. They are regarded as important in anaerobic fermentations; one of the lactate-utilisers frequently isolated from anaerobic reactor fluids is *Desulfovibrio* [3, 40, 41]. Other sulphate-reducing anaerobes are frequently reported in digesters: $3-5 \times 10^4$ per ml have been enumerated in digesting sewage sludge [42] and thermophilic sulphate reducers have also been found.

Lactate is a common end-product of bacterial fermentations and although many microbes form lactate as one of their end-products, the lactic acid bacteria produce lactate as their major end-product. This group consists of two broad categories, the homo- and the heterofermentative divisions, distinctions based upon the proportions of reaction products other than lactic acid formed during fermentation. Lactate production by the homofermentative strains may be summarized as:

$$\text{glucose} \rightarrow 2 \text{ lactate.} \tag{18}$$

The heterofermentative bacteria produce the following conversion:

$$\text{glucose} \rightarrow \text{lactate} + \text{ethanol} + CO_2. \tag{19}$$

Acetate can be produced instead of ethanol by some strains. The homofermentative division includes the streptococci and some lactobacilli; heterofermentative organisms include a number of lactobacilli and *Leuconostoc* species.

Leuconostoc mesenteroides, in common with a number of the lactic acid bacteria, can carry out the conversion of malic acid to lactate by at least two mechanisms. The first system comprises the decarboxylation of malic acid to pyruvate, a reaction mediated by an NAD-linked malic enzyme, followed by the reduction of the pyruvate to lactate. However, only L-malate is converted by this means, to L-lactate, a process requiring an L(+)-specific lactate dehydrogenase. This enzyme is not present in *L. mesenteroides*, which nonetheless is capable of this conversion. A second mechanism, therefore, dependent upon the possession of an enzyme system which directly converts L-malate to L-lactate without the intermediate formation of pyruvate and which needs NAD^+ as cofactor, is believed to exist [16, 43].

Some lactic acid bacteria e.g. *Streptococcus cremoris*, can use citrate as their carbon source to produce acetoin and diacetyl. This conversion is of importance in the treatment of milk-waste, which is high in citrate (milk contains approximately 1 g citrate l^{-1}). The typically anaerobic enzyme citrate lyase cleaves citrate to acetate and oxalacetate. The latter is decarboxylated to pyruvate and CO_2, and diacetyl is produced by subsequent dehydrogenation and condensation reactions. In summary:

$$3 \text{ citrate } \rightarrow \text{ lactate } + 3 \text{ acetate } + 5 CO_2 + \text{ diacetyl.} \qquad (20)$$

The diacetyl can be reduced to acetoin by an acetoin dehydrogenase in the lactate-forming bacteria, but the Enterobacteriaceae do not posses this enzyme and derive acetoin and butanediol by another pathway. In the latter group of organisms these end-product formations are dependent upon pH; above a pH value of 6.3 i.e. in the majority of anaerobic single-stage digesters at steady state, acetate and formate accumulate in the system and CO_2, H_2, acetoin and butanediol production by these microorganisms ceases.

2.4.2 The Bacteria of β-Oxidation

Acetate is the main fermentation product in the anaerobic digester under stable conditions but it can also be produced by the oxidation of fatty acids. In reactors treating sewage wastes, around 30% of the overall solids content of the digestible influent may be fats; in animal wastes, the fats comprise about 5–15% of the total solids [3]. The majority of the fats are triglycerides, i.e. glycerol substituted by fatty acids, and β-oxidation has been reported to be the mechanism of degradation of the fatty acids [31] to release an acetate molecule (C_2-unit) at each reaction. The cyclic pathway of β-oxidation is repeated until the saturated fatty acid is completely converted to C_2-units. Where a saturated fatty acid containing an odd number of carbon atoms is to be degraded, the last compound will be a C_3-unit or malonyl-CoA. This is an energetically unfavourable reaction and the presence of two bacteria is required. Although no evidence has yet been advanced to support the existence of a bacterium capable of degrading long chain fatty acids (e.g. stearic), an anaerobic microorganism that is able to convert the even-numbered-carbon fatty acids butyrate, caproate, and caprylate to acetate and H_2, and the odd-numbered-carbon fatty acids valerate and heptanoate to acetate, propionate and H_2, in coculture with either an H_2-utilising methanogen or an H_2-utilising Desulfovibrio has been reported [44]. The organism, which was present in numbers of at least 4.5×10^6 g^{-1} of anaerobic digester sludge, could only be grown in syntrophic association with the H_2-utilising bacterium, and no other sources of energy or combination of electron donors and acceptors were reported utilised. The degradation of the intermediate acid propionate to H_2 and acetate by *Syntrophobacter wolinii* in coculture with an H_2-utiliser has also been demonstrated [45], and propionate assimilation by at least three species of methanogenic bacteria was observed, the compound being incorporated almost exclusively into the C_2 of isoleucine [46].

Those microorganisms which produce acetic acid from propionate, butyrate and other higher fatty acids form the group known as the OHPA bacteria. The activity of these organisms is necessary to the anaerobic digestion process to degrade the longer-chain fatty acids, which cannot in their original states be utilised by the methanogenic bacteria [35]. The H_2 and acetate synthesised by the metabolism of the OHPA digester population have been estimated to provide the substrate for 54% of the total CH_4 produced in anaerobic reactor systems [47]. These bacteria are intolerant of pH fluctuations and must be maintained at neutral pH for maximum efficiency; their doubling times are of the order of 2 to 6 days [44, 45], i.e. longer in general than even those of the methanogens. The syntrophic association of OHPA and methanogenic bacteria allows the latter to assimilate the H_2 produced by the former.

The removal of the H_2 evolved by the OHPA bacteria is necessary for the degradation of fatty acids. Hydrogen-utilisation in a heterogeneous culture system such as an anaerobic digester pulls the fermentation processes from the production of reduced acids and ethanol towards the production of acetate and H_2 [3]. In consequence, acetate forms the main residual acid in a stable digester, its concentration depending upon the detention time of the system. High concentrations of propionate or butyrate are indicative of reactor failure, and propionate in particular is toxic to the H_2-utilising bacteria, so that system breakdown tends to be autocatalytic.

2.5 The Methanogenic Bacteria

Butyrate and propionate, as well as acetate, are thus converted to CH_4, the most reduced organic molecule. Methanogenesis is a conversion of H_2 and CO_2 by the CH_4-producing bacteria, generally considered to be the most oxygen-sensitive of the bacteria known and hence the most strictly anaerobic.

The family Methanobacteriaceae are structurally diverse and have been qualitatively divided into genera in the eighth edition of Bergey's Manual of Determinitive Bacteriology [48] on the basis of morphology. Species are characterised according to physiological and nutritional properties, but restricted correlation of inter-species relationships exists upon these bases, as the methanogens catabolise a limited range of substrates and other metabolic, molecular and biochemical characteristics have yet to be clearly elucidated. Deoxyribonucleic acid (DNA) base composition analysis has indicated that a wide variability exists in the moles percent of guanine and cytosine (G + C) of methanogen DNA. The known values range from 52% for *Methanobact. autotrophicum*, to 27.5% for *M. arbophilicum* [17]. An alternative approach to the inadequate taxonomic groupings above has been suggested [18]: relationships between various methanogenic species have been proposed, based upon the degrees of sequence homology found in the 16S rRNA fragments of various isolates, and evidence from investigations into cell wall structure and composition, lipid distribution and intermediary metabolism have supported these findings.

The isolation and cultivation of methanogenic bacteria depend upon procedures which produce strictly anaerobic conditions. Those techniques developed from the pioneering work of Hungate [49–51] have proved the most successful. Methanogens have also been presumptively identified by the use of fluorescence microscopy, which utilises the properties of the chromophores F_{420} and F_{342} [52, 53] found in all methanogenic species. F_{420} has been extracted by several techniques and fluorescence used to monitor potential methanogenic activity in anaerobic digester sludges [54, 55]. One rapid and sensitive method for the analysis of ^{14}C-labelled and unlabelled metabolic gases from methanogenic systems using chromatographic means has also been described [56].

Cytochromes of the b- and c-type, menaquinones and ubiquinones have not yet been detected in the methanogenic bacteria [17], and oxidative phosphorylation or substrate level phosphorylation mechanisms, if present, have not been identified. The methanogens cannot utilise complex organic compounds and their energy metabolism is directed towards a system that produces CH_4 as the only significant end-product. All methanogenic bacteria examined to date oxidise H_2 and reduce CO_2 to CH_4. Some species metabolise formate, while the members of *Methanosarcina* will also utilise methanol, methylamine (dimethyl-, trimethyl-, and ethyldimethylamine) and acetate as sole electron donor for growth and CH_4 production [18]. Methanogenesis by the reduction of CO_2 proceeds in a stepwise manner, but the intermediary products methanol, formate and formaldehyde remain carrier-bound (see Chap. 1).

The majority of methanogenic bacteria are autotrophic, but organic substrates such as acetate are necessary to the metabolism of some. Approximately 60% of the cell carbon of *Methanobact. ruminantium* derives from acetate; the transformation of acetic acid to CH_4 and CO_2 by *Methanosarcina barkeri* and *Methanospirillum hungatii* proceeds by the conversion of the methyl carbon of acetate, with its associated H-atoms, into methane: the carboxyl group provides the source of the CO_2. When CO is used as the sole energy source, the thermophile *Methanobact. thermoautotrophicum* produces 3 moles of CO_2 and 1 mole of CH_4 from 4 moles of CO, via the actions of a CO dehydrogenase and hydrogenase with the fluorescent carrier F_{420} as the electron acceptor [57]. The reduced F_{420} subsequently acts as the electron donor in the reaction reducing CO_2 to CH_4.

Conversion of acetate to CH_4 must be considered a rate-limiting reaction in digester design and operation as the acetate-assimilators are slow-growing. These organisms in a system digesting pig-waste had a maximum growth rate of about $0.4 \, d^{-1}$ at 35 °C [58]. The washout dilution rate of the acetate-utilising stage of a two phase digestion system was calculated to be $0.49 \, d^{-1}$ and a detention time of 100 h has been deemed suitable for the assimilation of fatty acids in the secondary stage of a laboratory two-staged digestion [59]. It has been suggested that acetate-utilising rods are the most numerous constituents of the digester flora [59] but in a digester treating piggery-waste, no acetate-utilisers were found, although counts of up to 2×10^6 methanogenic bacteria per ml were recorded and formate- and butyrate-utilisers appeared in equal proportions; domestic digester sludge was also reported to contain 2×10^2 formate-utilisers per ml of liquor [13].

Thermophilic methanogenesis from acetate by granular consortia of microorganisms (see Chap. 1) has been described and a working model for CH_4 formation from acetate presented [60]. The amalgamation of methanogens into consortia with other species not only aids the conversion reactions to CH_4 by enhancing

34

the proximity of the substrates and the removal of potentially toxic materials, but probably also provides a degree of protection from environmental fluctuations. Several thermophilic methanogens have been characterised [18, 61, 62].

2.6 Other Bacterial Conversions

The importance of sulphate-reducing microorganisms in the degradation of organic polymers to CH_4 in anaerobic digesters cannot be underestimated. Species of the genera *Desulfovibrio* and *Desulfotomaculum* are routinely isolated from digesters [63] and populations of 10^4 sulphate reducers per ml of reactor fluid have been recorded [42]. Complex carbon compounds can be catabolised by many of these species in the presence of excess sulphate, or in situations where sulphate is limited but where H_2-oxidising bacteria such as the methanogens are present [64]. The sulphate-reducing bacteria are known to grow via the coupled oxidation of immediate methanogen energy sources (i.e. acetate and H_2) to the reduction of sulphate [65, 66]. Complete oxidation of acetate to CO_2 under anaerobic conditions by *Desulfuromonas acetoxidans*, with elemental sulphur as the electron acceptor, has also been reported: the sulphur in this reaction is reduced to sulphide [67]. Thermophilic species of *Desulfovibrio* and *Desulfotomaculum* exist [68] and the presence of a thermophilic, sulphur-producing clostridial strain, *Cl. thermosulfurogenes*, has been reported [69].

The rather anomalous conversion of H_2 and CO_2 to acetate in the anaerobic digestion process has been postulated (see Fig. 10) and a spore-forming bacterium capable of this reaction has been isolated from sludge [70]. The utilisation of H_2 and CO_2 by sludge microorganisms to produce short-chain fatty acids, primarily acetate, was reported by other investigators [71, 72]; although the role of this reaction in the anaerobic biodegradation process has not been clarified. However, *Cl. aceticum* is an anaerobe capable of this conversion.

2.7 Anaerobiosis

One of the main reasons for the paucity of information on the various stages of anaerobic digestion is the difficulty in the cultivation of strictly anaerobic microorganisms; this difficulty can be traced to the extreme toxicity of oxygen to many anaerobic genera. Aerobic manipulation can be tolerated for short periods by some strict anaerobes (including methanogens) provided the culture media have been pre-reduced under anaerobic conditions [73]. Isolation procedures are being simplified due to the development of improved culture techniques, and microbial interactions within the anaerobic microcosm can be elucidated [51, 74] although the maintenance of anaerobiosis still remains the major impediment to research.

Anaerobic bacteria differ in their degrees of tolerance to oxygen, and hence to the Eh value of their environment. These tolerance ranges may be functions of the efficiencies of the microbial defence mechanisms against the toxic products of oxygen reduction. Molecular oxygen, upon interaction with various cellular constituents, is converted to the highly toxic hydrogen peroxide (H_2O_2) and the superoxide anion. When H_2O_2 and superoxide anions react intracellularly, hydroxyl radicals, the most potent oxidants known, are formed. Singlet oxygen, another cell-damaging moiety, is generated by the reaction of hydroxyl radicals and superoxide anions. The bacterial cell is protected against these potentially lethal reagents by the enzymes superoxide dismutase (SOD), catalase, and the peroxidases. The survival times of a number of anaerobic microorganisms under conditions of aerobiosis were investigated [75]: these times ranged from 45 min for *Peptostreptococcus anaerobius* to over 72 h for *Cl. perfringens* strains. The enzyme activity of several of the SOD-possessing strains of anaerobes was observed to correlate with aerotolerance, although this was by no means universal in the organisms investigated. None of the strict anaerobes examined possessed peroxidase activity, and variable catalase activity, unrelated to oxygen-tolerance, was recorded. However, enzyme activity of any description is dependent to a great extent on external conditions and also on the growth phase of the relevant bacteria. It has been suggested that the extreme oxygen sensitivity of methanogens during growth may reflect the significant oxygen-lability of key cofactors in the CH_4-generation pathway. The F_{420} moiety, during oxidation, has been particularly considered [76]. Enzyme denaturation is proposed to occur in an oxygenic environment, as enzymes when reduced are associated with F_{420} and are stable; enzymes when oxidized are dissociated from F_{420} and are labile.

The aim of the study of anaerobic microbiology in relation to anaerobic digestion systems is the improvement of the practical applications of the process to the treatment of industrial and other wastes. Examinations of the individual groups, the complex interrelationships and the many phases that comprise an anaerobic system provide an insight into the many mechanisms of anaerobiosis; several aspects require additional investigation for a fuller interpretation of the systems of anaerobic degradation, and their advantages when applied to the stabilisation of potentially hazardous wastewaters of environmental concern.

References

1. van Assche PF (1982) Antonie van Leeuwenhoek 48:520
2. Bryant MP (1979) J Animal Sci 48:193
3. Hobson PN (1982) Production of biogas from agricultural wastes. In: Subba Rao NS (ed) Advances in agricultural microbiology. Butterworth Scientific, London, p 523
4. Henze M, Harremoes P (1983) Water Sci Technol 15:1
5. Hills DJ (1979) Agric Wastes 1:267
6. Speece RE, McCarty PL (1964) Nutrient requirements and biological solids accumulation in anaerobic digestion. In: Avances in water pollution research: Proc of the Int Conference, Sept. 1962, London, vol 2. Pergamon Press, Oxford, p 305
7. van den Berg L, Lentz CP (1978) Food processing waste treatment by anaerobic digestion. In: Proc 32nd Ind Waste Conf, Purdue Univ, Lafayette, Indiana 1977. Ann Arbor Science, Ann Arbor, Michigan, p 252

8. Hoban DJ, van den Berg L (1979) J Appl Bacteriol 47:153
9. Lettinga G, van Velsen AFM, Hobma SW, de Zeeuw W, Klapwijk A (1980) Biotechnol Bioeng 22:699
10. Murray WD, van den Berg L (1981) Appl Environ Microbiol 42:502
11. Wijbenga DJ, Meiberg JMB, Brunt K (1984) Wastewater purification in the potato starch industry. In: Houwink EH, van der Meer RR (eds) Progress in industrial microbiology, vol 20: innovations in biotechnology. Elsevier Scientific Publishers, Amsterdam, p 121
12. Mah RA, Sussman C (1968) Appl Microbiol 16:358
13. Hobson PN, Shaw BG (1974) Water Res 8:507
14. Chin M (1983) Appl Environ Microbiol 45:1271
15. Hobson PN, Shaw BG (1973) Water Res 7:437
16. Doelle HW (1981) Basic metabolic processes. In: Rehm H-J, Reed G (eds) Biotechnology: a comprehensive treatise in 8 volumes, 1: microbial fundamentals. Verlag Chemie, Weinheim-Deerfield Beach, Florida Basel, p 113
17. Zeikus JG (1977) Microbial Rev 41:514
18. Balch WE, Fox GE, Magrum LJ, Woese CR, Wolfe RS (1979) Microbial Rev 43:260
19. Scott R (1979) Rennets and Cheese. In: Wiseman A (ed) Topics in enzyme and fermentation, biotechnology 3. Ellis Horwood, Chichester, p 101
20. Toerien DF (1967) Water Res 1:507
21. Siebert ML, Toerien DF (1969) Water Res 3:241
22. Fogarty WM, Kelly CT (1979) Developments in microbial extracellular enzymes. In: Wiseman A (ed) Topics in enzyme and fermentation biotechnology 3. Ellis Horwood, Chichester, p 45
23. Cooney CL, Wang DIC, Wang S-D, Gordon J, Jiminez M (1979) Biotechnol Bioeng Symp 8:103
24. Lee BH, Blackburn TH (1975) Appl Microbiol 30:346
25. Khan AW, Trottier TM (1978) Appl Environ Microbiol 35:1027
26. Kahn AW, Trottier TM, Patel GB, Martin SM (1979) J Gen Microbiol 112:365
27. Khan AW, Mes-Hartree M (1981) Appl Microbiol 50:283
28. Latham MJ, Wolin MJ (1977) Appl Environ Microbiol 34:297
29. Weimer PJ, Zeikus JG (1977) Appl Environ Microbiol 33:289
30. Eastman JA, Ferguson JF (1977) J Water Pollut Control Fed 53:352
31. Gujer W, Zehnder AJB (1983) Water Sci Technol 15:127
32. Toerien DF, Hattingh WHJ (1969) Water Res 3:385
33. Sorensen J, Christiensen D, Jorgensen BB (1981) Appl Environ Microbiol 42:5
34. Widdell F, Pfennig N (1982) Arch Microbiol 131:360
35. Verstraete W, de Baere L, Rozzi A (1981) Trib Cebedeau 34:367
36. Nagase M, Matsuo T (1982) Biotechnol Bioeng 24:2227
37. Bauchop T (1967) J Bacteriol 94:171
38. Theil PG (1969) Water Res 3:215
39. Chung KT (1976) Appl Environ Microbiol 31:342
40. Ueki A, Minato H, Azuma R, Suto T (1980) J Gen Appl Microbiol 26:25
41. Ueki A, Suto T (1981) J Gen Appl Microbiol 27:229
42. Toerien DF, Thiel PG, Hattingh WHJ (1968) Water Res 2:505
43. Kunkee RE (1974) Adv Chem Ser 137:151
44. McInerney MJ, Bryant MP, Pfennig N (1979) Arch Microbiol 122:129
45. Boon DR, Bryant MP (1980) Appl Environ Microbiol 40:626
46. Eikmanns R, Jaenchen R, Thauer RK (1983) Arch Microbiol 136:106
47. Kaspar HF, Wuhrmann K (1978) Appl Environ Microbiol 36:1
48. Bryant MP (1974[8]) Methane producing bacteria, part 13 of: Buchanan RE, Gibbons NE, Cowan ST, Holt JG, Liston J, Murray RGE, Niven CF, Ravin AW, Stanier RY (eds) Bergey's manual of determinitive bacteriology. Williams and Wilkins, Baltimore, p 472
49. Hungate RE (1950) Bacteriol Rev 14:1
50. Hungate RE (1969) A roll-tube method for cultivation of strict anaerobes. In: Norris JR, Ribbons DW (eds) Methods in microbiology, vol 3 B. Academic Press, New York, p 117
51. Latham MJ, Wolin MJ (1978) Use of a serum bottle technique to study interactions between strict anaerobes in mixed cultures. In: Lovelock DW, Davis R (eds) Techniques for the study of mixed populations. Academic Press, London, p 113
52. Mink RW, Dugan PR (1977) Appl Environ Microbiol 27:985

53. Doddema HJ, Vogels GD (1978) Appl Environ Microbiol 36:752
54. Delafontaine MJ, Naveau HP, Nyns EJ (1979) Biotechnol Lett 1:71
55. Binot RA, Naveau HP, Nyns EJ (1981) Biotechnol Lett 3:632
56. Nelson DR, Zeikus JG (1974) Appl Microbiol 28:258
57. Daniels L, Fuchs G, Thauer RK, Zeikus JG (1977) J Bacteriol 132:118
58. Hobson PN, McDonald I (1980) J Chem Technol Biotechnol 30:405
59. Cohen A, Zoetemeyer RJ, van Deursen A, van Andel JG (1979) Water Res 13:571
60. Bochem HP, Schoberth SM, Sprey B, Wengher P (1982) Can J Microbiol 28:500
61. Stetter KO, Thomm M, Winter G, Wildgruber G, Huber H, Zillig W, Janekovic D, Konig H, Palm P, Wunderl S (1981) Z Bakt Parasit Infekt Hyg 1:166
62. Rivard CJ, Smith PH (1982) Int J Syst Bacteriol 32:430
63. Zeikus JG (1979) Microbial populations in digesters. In: Proc 1st Int Symp on Anaerobic Digestion, Sept 1979 Cardiff UK. Applied Science Publications Ltd, London, p 61
64. Bryant MP, Campbell LC, Reddy C, Crabill M (1972) Appl Environ Microbiol 33:1162
65. Widdell W, Pfennig N (1977) Arch Microbiol 112:119
66. Badziog W, Thauer R, Zeikus JG (1978) Arch Microbiol 116:41
67. Pfennig N, Bieble H (1976) Arch Microbiol 110:3
68. Zeikus JG (1979) Enzyme Microb Technol 1:243
69. Schink B, Zeikus JG (1983) J Gen Microbiol 129:1149
70. Ohwaki K, Hungate RE (1977) Appl Environ Microbiol 33:1270
71. Genther BRS, Davis CL, Bryant MP (1981) Appl Environ Microbiol 42:12
72. Goldberg I, Cooney CL (1981) Appl Environ Microbiol 41:148
73. Patel GB, Roth LA, Agnew BJ (1984) Can J Microbiol 30:228
74. Bhatnagar L, Henriquet M, Longin R (1983) Biotechnol Lett 5:39
75. Rolfe RD, Hentges DJ, Campbell BJ, Barrett JT (1978) Appl Environ Microbiol 36:306
76. Cheeseman P, Toms-Wood A, Wolfe RS (1972) J Bacteriol 112:527

3 Forms of Biomass

Microbial cells exist in a range of sizes, shapes and phases of growth, individually or aggregated into various microstructures. These conditions are of practical significance in anaerobic digestion since the form of the biomass is likely to have a significant effect on organism survival and nutrient transfer, and thus the efficiency of the overall digestion process. In a turbulent system, attached biomass can persist whilst cells in suspension are lost with the effluent [1]. Abiotic suspended particles may be utilised as adhesion sites for bacteria, aiding their persistence by enhanced sedimentation and hence their avoidance of washout in the effluent. Microstructural forms of biomass are shown in Fig. 11; these can be major determinants of mass transfer. Formation of a particular structural aggregate depends on several factors including the size range of cells within the microbial population and the location of each individual cell relative to other cells and the medium, for example at a gas/liquid interface. Non-uniform gradients of organic compounds, ions, enzymes and conductivity (due to bacterial metabolism) exist as the aggregates are non-homogeneous, filamentous forms sometimes predominating.

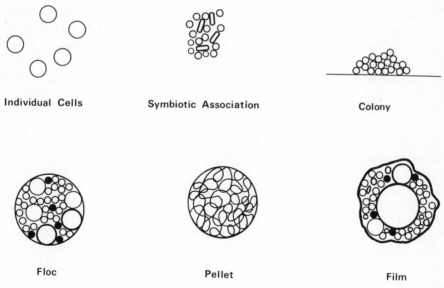

Fig. 11. Forms of biomass

3.1 Adhesion

Microbial habitats in aqueous systems such as anaerobic digesters are diverse, and survival and growth will depend upon factors such as temperature, nutrient fluctuation and stratification. In many instances organisms overcome environmental instability by adhesion to surfaces by hold-fasts, fimbriae and/or capsular polysaccharide, or by microcolony formation. However, relatively little is known about the mechanisms of attachment to surfaces under anaerobic reactor conditions.

The adhesive capacities of bacteria are exceptional; their surface structures appear to allow some control over adhesion while their small dimensions ensure that they are not subject to many naturally-occurring shear forces. The outer layers of bacteria are negatively charged, as are most of the surfaces to which they adhere. The range and intensity of the repulsive forces generated by these surface charges as the microorganism and the surface converge will be governed by the ionic strength of the medium, the energy level of the cell, pH, and any potentially adsorbing ions present; the repulsion will probably not reduce to zero under most conditions [2]. Estimation of the van der Waals attractive force between the surface and the organism is difficult. This force is electromagnetic in origin and can be described using A, the Hamaker constant, which represents all the molecular interactions between the adjacent surfaces. The van der Waals energy of two interacting flat surfaces is described by the following equation [3]:

$$E_a = -A/12\pi d^2, \tag{21}$$

where E_a is the attractive van der Waals-type energy per unit area, and d is the distance between the two surfaces. Values of A that have been used are only estimates at best, and evaluations of the resultant degrees of attraction or repulsion at close range are hence probably ineffectual, as these are dependent on the Hamaker constant [2]. As microbial cells can be removed from surfaces by sonication or shear without disrupting the cell membrane, Rutter [2] suggests that some residual repulsion does occur between bacteria and collecting surfaces.

The hydrophilic polysaccharide glycocalyx of some microbial cells enhances adhesion to non-active surfaces or to other cells. In an environment where attachment is of little value, the energy expended in production of the glycocalyx is effectively wasted, and therefore selection of mutants or strains without the polymeric coating may occur. Laboratory culture techniques tend to select for these strains and hence information on the mechanisms of adhesion has been slow to accumulate.

Adhesion of cells to a surface is complicated by the presence of water, as the bond forming in such an aqueous environment must resist hydrolysis. Microbial cells are readily adsorbed onto materials with hydrophilic surfaces, but hydrophobic solids such as Teflon and polyethylene are resistant to microbial attachment; those cells not strongly adsorbed are readily sloughed off [1]. Surface roughness also influences microbial cell attachment, by providing more shelter for the small particles from shear forces, and by increasing the available surface area for attachment.

The ionic binding of microbial cells to a surface requires the participation of the counter ions in the Stern layer adjacent to the surface. As the new bonds form, a rearrangement of the ions and charges maintains electrical neutrality. Hydrogen and hydroxyl ions generally have major roles at surfaces; pH is therefore a dominant factor in adhesion. Although ionic bonding may be important in the initial attachment steps of adhesion, stronger bonds are required for fixing microbial communities. Hydrogen bonding provides strong attachment forces, for although individual hydrogen bonds are weak, they exist in such great numbers that the net bonding strength is considerable. Adsorption of functional groups to the surface provides additional holding forces. However, when an attached aggregate increases in size and is sheared or sloughed off the substratum, an exposed surface is usually revealed, i.e. the internal bonds of the structure appear to be stronger than the adhesive forces allowing attachment to the surface [1].

Divalent cations, in particular calcium (Ca^{2+}) are reported necessary for adhesion of cells, although the strontium ion has been found to act as a less efficient alternative [4]. These divalent ions may stabilise the glycocalyx structure, or act as linkages between negatively charged surfaces and microorganisms, or they may enhance precipitation of the bacterial extracellular polysaccharide when the cell approaches closely to a surface [2].

It is generally believed that microbial activity is enhanced at surfaces, especially in environments of low nutrient status. This phenomenon has not been clearly elucidated but an increase in the surface area to volume ratio may enhance activity for several reasons. The most commonly accepted explanation is that the adsorption and concentration of nutrients or growth factors at surfaces increases the availability of these substances to the attached microorganisms [5]. However, although potential nutrients may be concentrated at interfaces, bacterial uptake of adsorbed substances may not be of consequence. The more efficient use of exoenzymes at surfaces has also been suggested [6]: enzyme adsorption may cause advantageous configurational changes or slow down enzyme diffusion from the surface. If surfaces can remove inhibitors from the environment, bacterial growth may also be facilitated. Finally, the solid/liquid boundary will be subjected to a range of physical and chemical factors different to those in operation in the bulk of the medium. These are predominantly related to the intermolecular forces between the two phases and the interfacial energy, and include pH and redox potential.

The net rate of microbial cell adhesion is the difference between the rate of cell adhesion and the rate of cell detachment. Adhesion is generally considered to be a two-stage process, where initial reversible adhesion is followed by an irreversible step, where permanent bonding occurs. This is usually aided by the production of extracellular polymeric substances. Removal after this latter stage is generally only possible by severe mechanical or chemical treatment: cell detachment may occur as the result of:

1) fluid dynamic forces;
2) shear forces;

3) lift; or
4) taxis [4],

factors all in operation in attached-biomass anaerobic reactors.

There are two generally accepted theories that are fundamentally concerned with the initial interactions of microorganisms and substrata: these are the DVLO Theory of Adhesion (named after Derjaguin and Landau [7] and Verwey and Overbeek [8]), where the electrostatic properties and van der Waals forces of the system are considered, and the "Wettability" or Interfacial Free Energy System Theory.

3.1.1 The DVLO Theory

This was generally considered a major landmark in colloid science. In the system, the positions of attraction have been named the primary minimum (at extremely small separation distances) and the secondary minimum (typically at 5–8 nm distances). The forces of repulsion reach a maximum between these two positions. However, the values used for the surface charges, the three-dimensional configuration of the attachment site and the varying dielectric constant of the liquid medium as the two surfaces come together, result in problems with the theory, despite the fact that the secondary minimum is amenable to relatively accurate theoretical prediction. Additionally, in these types of system, the measurement of Hamaker's constant is not possible [2]. The prediction by the DVLO theory that reversible adhesion can take place at the secondary minimum has received confirmation [4]. Theoretically, specificity of attachment at either the primary or the secondary minimum may be accounted for by the consideration of van der Waals and electrostatic forces in isolation; other adhesive forces such as polymer bridging may however come into effect at these distances. It is nonetheless improbable that bacterial cells can approach to within less than 1 nm of a substratum, and thus overcome the repulsive peak existing between the primary and secondary minima [4]. Bacterial fimbriae are frequently about 4 to 7 nm in diameter, and up to 1000 nm in length; the usual arrangement is peritrichous. Fimbriae are therefore thin enough to evade the energy minima between cell and surface, and so organised that the relative position of the bacterium to the surface is unimportant [9]. The energy barrier may also be breached by the extrusion of extracellular polymeric material by the cell; the polymeric molecules will likewise be able to approach the surface.

3.1.2 Interfacial Free Energy and Adhesion

Interfacial free energy, or tension, can be considered as the free energy of interaction between two media across a third in the limit of zero separation. In theory, if the contact of a microbial cell and an adjacent surface reduces the amount of total free energy in the system, then adhesion of the cell to the substratum will occur. Cell adhesion has been related to the critical wetting tension [10] although

42

Pethica [11] found the relationship to be no more than qualitative. The existing theories of force do not allow the zero separation limit to be predicted with any degree of certainty; theories of the local organisation of water molecules are not sufficiently complex and, primarily for this reason, the prediction of surface tension and interfacial free energy is not possible [12]. The dependence of interfacial energy upon temperature can be inconsistent with that of van der Waals forces, but the concepts of hydrophilicity, wettability and contact angle are useful in the consideration of adhesion.

3.1.3 Deformation in Relation to Adhesion

When contact between solids results in adhesion, deformation at site of contact is to be expected. The extent and pattern of deformation depends upon the conformation and elastic properties of the contacting materials and the forces between them [13, 14]. More extensive deformation is caused in the softer, less elastic solid by the interaction forces between closely adjacent solids [15]. Elastic energy is initially stored by the deformed solid; this storage is permanent if the solid is perfectly elastic and only dissipated when separation occurs. Plastic deformation results when the solid is viscoelastic, and elastic energy can be dissipated while the solids remain contiguous. Major differences in contact strength between the solids are manifested by these two types. Solids such as cells, which undergo plastic deformation on contact with surfaces, are more strongly attached, as the net adhesive force between elastic solids at a specific separation distance between the surfaces approximates to that value for the case of zero deformation [14]. When in contact with surfaces which give rise to highly dynamic attachments (e.g. metal) bacteria deform on collision. To maintain essential metabolic functions however, the avoidance of extreme contact deformation is likely to be a necessary characteristic of viable adherent microorganisms [12]. In this regard, attractive dispersion forces are counterbalanced in particular by electrostatic forces or solvent-mediated forces [12], thereby lessening deformation pressures and ensuring maximal stability of the internal structure of attached microorganisms.

Thus, the ability to adhere to surfaces provides an important microbial survival mechanism in relatively extreme environments such as anaerobic bioreactors, in particular in the packed bed and expanded and fluidised bed systems. Investigations into the mechanisms of adherence must provide some insight into the conditions within such reactors.

3.2 Biofilm Formation

Almost any surface immersed in a fluid medium can be colonised by bacterial cells. The attached organisms grow and proliferate, and may manufacture extracellular polymeric substances which form a structural matrix around the cell and beyond, eventually producing a structured mat of interlinked fibres or biofilm.

44

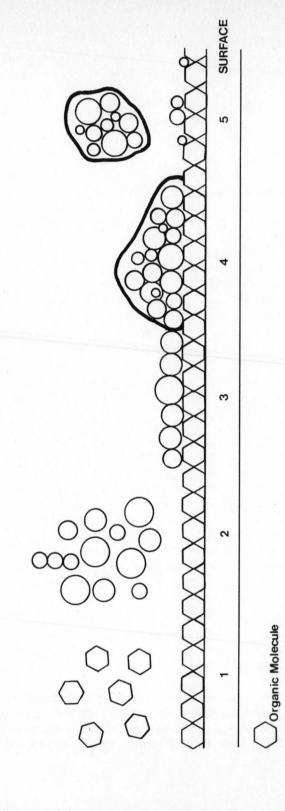

Direction of Fluid Flow

SURFACE

1 2 3 4 5

◯ Organic Molecule

◯ Bacterium

Fig. 12. Formation of biofilm. (After Trulear and Characklis [16])

The generation of the biofilm on a surface under conditions of fluid flow is the result of several biological processes, primarily:

1) transport of organic molecules, and their adsorption to the surface;
2) transport of microorganisms to the surface;
3) attachment of the bacteria by the recognised two-stage adhesion process, i.e.
   (i) rapid initial approach of cell to surface, influenced by electrostatic repulsion at long range and van der Waals attraction at short range, and
   (ii) firm attachment of cells by polymer bridging and steric interactions, a slower process;
4) production of the biofilm by the metabolic activity of the attached microorganisms; and
5) detachment of areas of the mature biofilm by hydraulic forces (see Fig. 12).

When an initially pristine surface such as an inert particle or a reactor wall is exposed to a fluid flow containing suspended organic molecules, nutrients and microbial cells, adsorption of an organic monolayer is frequently described as almost instantaneous [4]. This adsorbed layer conditions the surface of the stratum prior to attachment of the microorganisms. Bacterial transport to surfaces is regulated by aspects of fluid flow: the Brownian effect, together with gravitational and dynamic forces, may take effect where bulk fluid flow is absent; under turbulent flow conditions the above forces are negligible and eddy current effects and molecular diffusion are likely to be dominant [16].

The participation of extracellular structures such as fimbriae and holdfasts in biofilm formation has not generally been confirmed. Fimbriae are found on some members of the Enterobacteriaceae and the Pseudomonadaceae, and although the adhesion of the enterobacterium *Escherichia coli* to mammalian epithelial cells requires the presence of fimbriae they are not involved in the adhesion of the bacterium to glass. Some prosthecate bacteria, such as the caulobacters, produce holdfasts and inorganic cements are deposited by some species. The fimbriate human oral bacteria *Actinomyces* and *Bacteroides* are major participants in the formation of dental plaque, a form of biofilm. Bacterial flagella, although probably involved in the mobilisation of cells to the secondary minimum, have not been clearly implicated in biofilm adhesion. Biofilm formation appears to be primarily mediated by the extracellular polymeric material that comprises the capsular glycocalyces of almost all bacteria in natural environments, the very highly hydrated (99%) fibrous polysaccharide matrix constituting the basic substance of adherent biofilms [17]. Within the organic matrix however, in addition to the polysaccharides and sugars defined by Costerton [17], other macromolecules, including glycoproteins [18] proteins and nucleic acids [19, 20] have been reported. A mixed bacterial population derived from a bioreactor was found to contain polysaccharides, proteins and nucleic acids in the approximate ratio 3:2:2 [21].

The glycocalyx material is in essence an ion exchange resin [22] and the biofilm configuration is important in the protection of bacteria from antibacterial agents. Electron microscopy has shown a large number of sessile microorganisms entrapped in a fibrillar anionic superstructure, forming a biofilm that is adherent to the substratum [17]. On a nutrient substratum (bitumen), the bacterial multiplication in the biofilm produced digested pits on its surface. The biofilm mode of growth permits the aggregation of the microorganisms into consortia and the microbial enzymes are concentrated near the cells by binding to the film matrix [17]. According to Trulear and Characklis [16] preferential selection of bacterial species for the biofilm probably occurs; this selection is dependent upon the level of influent nutrient (glucose). At low glucose loading rates the high surface area to volume ratios of filamentous bacteria and their extension into the aqueous phase provided them with a physiological advantage over the other microbial forms present in the system.

Biofilms are produced on a wide variety of plastic, metal, ceramic and other components upon which the bacterial cells are typically enmeshed in the condensed remnants of their exopolysaccharide glycocalyces. Thick, adherent biofilms have been shown to protect their indigenous microflora from potentially harmful materials such as antibiotics at concentrations many times higher than planktonic in vitro Minimum Inhibitory Concentration (MIC) levels [23]. Detailed studies in pathogenic and industrial systems of glycocalyx-enclosed bacteria have indicated that the cells are protected to a marked extent because antibiotic and chemical agents cannot penetrate the anionic matrix of the enveloping glycocalyces [24, 25].

According to Characklis and Cooksey [4], the extracellular polymeric material may aid microbial film processes in various other ways, including:

1) strengthening of bonding and other adhesive forces;
2) protection of entrapped cells from rapid changes in the nature of the influent wastestream;
3) adsorption of substrates;
4) adsorption of toxic materials from the system;
5) provision of short term energy storage via cell membrane potential; and,
6) enhancement of the intercellular transfer of genetic material.

Physical biofilm properties including thermal conductivity and diffusibility may also be influenced by the nature of the matrix.

Detachment processes have been referred to as shearing or sloughing; the former describes the continuous, fluid-dynamic dependent removal of small areas of the biofilm whereas sloughing refers to a random, bulk film detachment, usually considered to be the result of a depletion of nutrient substances deep within the biofilm [26] or some significant environmental change. Detachment may be the outcome of hydrodynamic [27] or chemical influences.

Thus, physical, chemical and biological biofilm characteristics are functions of the environmental conditions surrounding the attachment surface (e.g., the interior of an anaerobic digester) and the prevalent organisms are selected for by a combination of the physical and chemical components of the microenvironment of the surface. As the biofilm evolves, gradients develop within it and its inherent properties alter. The physical properties include the thermodynamic parameters of volume (thickness) and mass, and transport properties [4]. Biofilm density increases with increase of substrate loading and turbulence [16], the latter probably caused by one of the following:

1) the effects on microbial metabolism of environmental stresses;

2) selective adhesion of particular bacterial species from those populations present in the system; and
3) forcing out of loosely bound water from the biofilm by fluid pressures [4].

Both inorganic and organic chemical properties affect biofilm characteristics. The former varies with the chemical composition of the influent medium, and the latter with the energy and carbon sources available. Biofilm development rate and chemical and physical properties are influenced by the colonising microbes. Interactions between cells, and also between cells and environment, cause a successional effect during biofilm evolution. Biofilms thus produce the following results in aquatic systems:

1) they increase fluid frictional resistance by:
   (i) increasing particle surface roughness, an effect enhanced by filamentous organisms, and
   (ii) increasing drag via their viscoelastic properties;
2) they impede conduction and convection of heat through their matrices.

The latter effect may decrease the sensitivity of the system to temperature fluctuation, an important consideration in anaerobic treatment processes.

Rather than a fixed surface film, the controlled pore ceramic system of Messing [28] used a three-dimensional film structure, with cells immobilised on the internal walls of the carrier media. When the cells were immobilised on their minor dimension a maximal microbial surface area for nutrient transfer and waste removal resulted. One phenomenon noted during this process was that the large number of *E. coli* K12 cells present in the effluent had a major dimension of 0.65–0.85 µm, while the culture immobilised in the first instance comprised cells of 1–6 µm in length [29]. The authors observed that the maintenance of these so-called "mini-cells" in medium for periods in excess of one hour resulted in a return to normal cell length, i.e. 1–6 µm. "Mini-cell" formation was attributed therefore to early cell division, a consequence of the rapid influx of nutrients and removal of metabolic wastes.

As biofilm-entrapped microbes, particularly in the expanded and fluidised bed anaerobic systems, are retained in numbers sufficient to allow a significant reduction in retention time within the bioreactor, the ability to process wastewater on an efficient and continuous basis has rendered the use of anaerobic microorganisms in biological reactor systems much more attractive.

3.3 Floc Formation

Flocculation is of practical significance in the treatment of wastes, as flocculated microstructures are easily collectable by sedimentation from separated streams in a wastewater treatment plant. The phenomenon is of particular importance in the contact and carrier-assisted contact reactor processes, and also in the upflow anaerobic sludge blanket reactor system. Flocculant growth is not necessary for efficient substrate removal [30] but is essential for ensuring clear effluent. The formation of flocculated structures can occur by several means and although controversy exists as to the actual mechanisms operating, the general consensus of opinion points to elucidation of the phenomenon by the concepts of colloid science.

Most colloidal particles in natural aquatic systems carry a net negative charge, and hence repel one another. This repulsion therefore prevents their coalescence into particles large enough to precipitate. The net negative charge is estimated as the zeta potential and is defined as the potential difference between the fixed layer of ions on the particle surface and a point in the aquatic medium where electroneutrality exists. In general, coagulation refers to the neutralisation of the surface charge on the particles. Agitation of the medium results in the formation of microflocs, by increasing the collision frequency of the small neutralised particles. Further dimensional enlargement occurs into structures of sufficient size to settle rapidly, in a sedimentation basin for instance. This growth, caused by externally applied movement, is known as flocculation. However, where mass fluid motion is not present, both Brownian diffusion and gravity-induced flocculation (differential creaming) can cause flocculation of polydisperse collodial particles if the particle-particle interactions are favourable. Gravity induced flocculation is the result of the differential sedimentation rates between large and small particles, where the faster-creaming large particles eventually sweep out the slower-creaming particles in their path, with flocculation possibly occurring [31, 32], due to effects including convection currents, shear fields and particle-particle interactions. If particles are not spherical in shape and have variable surface charges, discrete sedimentation is hampered as it is a function only of settling velocity. Increases in size, shape and density can occur during flocculant (indiscrete) sedimentation, where removal of particles from the aqueous phase is a function of both settling velocity and time [33]. Flocculant supensions contain particles which continue to increase in size during the sedimentation process.

There is an important distinction between flocculant growth, which is the non-separation of daughter cells from parents, and flocculation – the aggregation of dispersed cells. As two cells come together from the direction of infinite separation, they will aggregate if they possess the kinetic energy with which to override any existing energy barrier; the cells remain together if the energy barrier at zero separation cannot be breached. As a result, the disruption of flocculated microstructures by shearing forces may not be reversed by a return to latent environmental conditions. In addition, microbial growth as flocs or discrete cells cannot give an indication of the presence or absence of an energetic impediment to flocculation, but does suggest the depth of the primary minimum. The latter is probably often shallow, as evidenced by the great number of microorganisms that grow singly. Thus, a strengthening of the attractive forces in the primary minimum is required to flocculate such cells, and possibly also the depression of an energy barrier antagonising aggregation [34].

Collision frequency is a factor in flocculation kinetics. For microorganisms of diameter greater than 1 μm, moderate shear stresses in the system caused by e.g. moderate agitation or thermophoretic currents significantly increase collision frequency. Motility can also be of consequence. Hence, as an indication of the flocculation characteristics of microorganisms, rates of flocculation used must be estimated in relation to specified temperature, particle density, shear and motility conditions [34].

Most particles suspended in water and wastewater such as bacterial cells, silica, clays, paper fibres or hydrated metal oxides possess negative surface charges at neutral pH. According to Daniels [33], surface charges on suspended particles are primarily produced by one or more of three distinct processes, depending on the composition of the solution and/or the nature of the solid phase:

1) the variety of surface groups capable of ionisation e.g., amino, carboxyl, hydroxyl, sulphato. The charge on these particles is dependent upon solution pH and the extent of surface group ionisation;
2) preferential adsorption of certain ions from solution ("potential determining ions") which can occur via covalent or hydrogen or van der Waals bonding and can be intensified by electrostatic attraction; and
3) imperfections in a crystal lattice – this source accounts for a major portion of the charge on clay minerals; the aqueous phase has no effect on the size and magnitude of the charge produced by isomorphic replacements of individual atoms.

The role of surface charge in flocculation has not been clearly elucidated. Sludge settleability in activated sludge systems was reported to be related to the surface charges of sludge particles, as well as those of the component microorganisms [35, 36]. Reduction of surface potential however, was not found to be a principal determinant of flocculation [21].

In addition to the fixed cell-surface charges, the protonmotive force or energy-generating system of the cell, defined by the equation:

$$\Delta p = \Delta \psi - Z \Delta pH \tag{22}$$

has also been implicated in flocculation [37]. In the above equation Δp is the protonmotive force (p.m.f.) in mV, $\Delta \psi$ is the membrane potential in mV, and ΔpH is the transmembrane pH gradient; Z is the factor that converts pH into mV (59 at 25 °C) [38]. The p.m.f. and its components are involved in the regulation of several aspects of cellular metabolism, including the translocation of cations and anions across membranes [39].

The formation of an aggregated floc particle occurs when one end of a flocculant organism attaches to the surface of another at one or more adsorption sites and the other extruding, unadsorbed end of the first organism bridges and adsorbs to one or more additional particles. Four major factors favour the aggregation of bacteria [40]. These are listed in Table 5.

Table 5. Major factors favouring bacterial aggregation. (After Ash [34])

Gel entrapment	Viscous effect
Incipient flocculation	Interaction between adsorbed or attached microorganisms – short range forces
Charge-mosaic interaction	Flocculation by cationic polyelectrolytes
Polymer bridging	Flocculation by polymer addition – short range forces

One mechanism, independent of the thermodynamic considerations which influence dispersion as the stable state of the system rather than aggregation, is the enmeshment of microorganisms in a gel of extracellular polysaccharide (see biofilm formation, above). The elevated matrix viscosity reduces the rate at which a stable dispersion could be formed [34].

Adsorbed or attached large molecules tend to sterically stabilise microbial cells, because overlap of the polymer layer is energetically disadvantageous, tending to increase system free energy. When the polymer layers overlap and the resultant free energy is negative, then incipient flocculation results [41]. Ash [34] considers this effect in relation to short range bonding between polymer segments. Hydrogen, ion pair and triple ion bonding appear to be the most significant of the short range interactions in biopolymers [42]. Conditions which increase the number and strength of these linkages will enhance floc formation: one such factor is the presence of nonbiological polymeric material in the system.

Bacterial aggregation during the endogenous growth phase in microbial systems has been attributed to the products of lysis [43, 44], and this has been considered a factor in sludge flocculation [37]. The proteinaceous nutrient waste of some food-manufacturing processes, such as those of the bakery and soya-processing industries [34] may also add to the amount of polymeric material in the reactor, and macromolecules produced by the protozoa present in some digesters such as the expanded bed [45] can likewise increase the polymer present [46].

Cationic polyelectrolytes may bring about flocculation of the negatively charged microorganisms by reducing the net surface charge or by introducing interactions between charge mosaics [40]. At polyelectrolyte molecule adsorption sites the negative surface charges of the cell are neutralised. The adsorbed polyelectrolytes also cause areas of charge reversal on the microbial cell surface because of the excess cationic segments. Strong electrostatic attractive forces come into play when these localised regions of positive charge on the bacterial cell align with the negative areas on an approaching organism, and double-layer interaction energies are reduced. However, excess polymer leads,to steric or electrostatic repulsion. Nonmicrobiological polymeric materials may comprise part of some industrial effluents and may therefore be components of wastestreams undergoing treatment.

The fourth mechanism of flocculation enhancement is polymer bridging, which occurs when a polymer is added to a suspension of microorganisms; the phenomenon is defined as the adsorption of a polymer molecule onto more than one particle surface simultaneously. The dynamic forces that are necessary to incipient flocculation (above) are disadvantageous for flocculation by polymer bridge, as bridge formation is benefitted by the extended conformation of the molecule [34]. Frequently, the distinction between incipient flocculation and polymer bridging in microbial systems is ill-applied, since a lack of information on the state of the polymers at the surface of microorganisms exists.

Although microbial aggregation has been extensively studied, the elucidation of the mechanisms operating, within the sphere of colloid theory, is difficult because of the inadequacy of relevant data. However, observations suggest the participation of a range of mechanisms.

3.4 Pellet Formation

The phenomenon of microbial pellet or granule formation has been recorded in many instances [47–50], but in the treatment of wastewater, granulation appears to be restricted to UASB reactor types, and generally to those treating carbohydrate-rich or VFA-type wastes [50]. Films and flocs are continuous but pellets may be networks of filaments (see Fig. 11). An examination of the bacterial morphology of pellets from a full scale upflow anaerobic clarigester by scanning electron microscopy revealed rod-like and filamentous microbial forms and also the presence of extracellular polymer [50]. The pellets were observed to be held intact by fibrous strands interwoven with the component granule bacteria; the network of strands was generally composed of polysaccharide.

Fluid can flow though microbial pellets, but only at a slow rate, as the open spaces within the structure have high resistance due to their size. Viscosity of film retards convection and slows molecular diffusion, whereas hydraulic deformation of a filamentous granule may force out interstitial water and thus aid mass transfer [1].

The mechanisms of granulation have not been fully defined, but environmental conditions seem to be important [51]. These include temperature, pH (6.5–6.8), accessibility of essential elements and the composition of the wastewater. The latter includes in particular the degree of biodegradability of the component organic matter, the presence and level of inhibitors, and the properties of discrete nonbiodegradable materials of both organic and inorganic origin, such as mud solids. The species of seeding sludge and digester start-up procedures including loading rate and amount of seed sludge are also important.

Pellet formation has been reported mainly for methanogenic sludges, but granulation has also been observed in an UASB reactor operating in a denitrifying process [52]. The pellets were formed at high loading rates and were initially of a uniform size (2–4 mm diameter) but later in the operation, large (5 mm) and small (0.5 mm) pellets were found. The formation of various types of granules in different reactors may occur by different mechanisms; larger pellets provide a low surface area, and substrate penetration rates may be poorer. Pelletisation is an active process which cannot at present be quantified; the onset of its formation is unclear. Hulsoff-Pol et al. [51] have distinguished three phases:

1) expansion of sludge bed due to filamentous growth and onset of gas production;
2) formation of distinct granules in the heavier fraction of retained sludge in the lower region of the reactor; and
3) concentrated granular growth exceeding sludge washout until a period of stagnation occurs.

Different reactor loading rates appear to produce varied granule forms. At lower rates (4 kg COD m^{-3}d^{-1}), the pellets were observed to be filamentous, and composed of long polycellular filaments of rod-shaped organisms [51]; inert inclusions originating from the media have also been reported [51, 52]. At higher loading rates (5.5–6.0 kg COD m^{-3}d^{-1}), short multicellular fragments consisting of

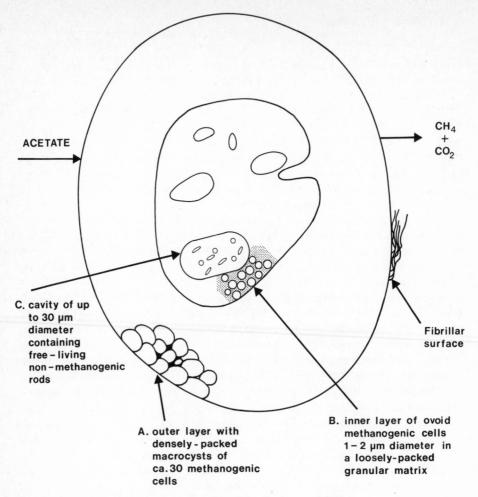

ACETATE

CH$_4$
+
CO$_2$

C. cavity of up
to 30 µm
diameter
containing
free – living
non – methanogenic
rods

Fibrillar
surface

A. outer layer with
densely - packed
macrocysts of
ca. 30 methanogenic
cells

B. inner layer of ovoid
methanogenic cells
1 – 2 µm diameter in
a loosely-packed
granular matrix

Fig. 13. Representation of a granular consortium. (After Bochem et al. [53])

approximately four cells were observed [51], which apparently have a lower activity than the filamentous type of pellet.

The morphology and ultrastructure of thermophilic granules have been investigated by light and electron microscopy [53]. Three distinct strata could be visualized in pellets of 1–3 mm in diameter (Fig. 13): A, a densely packed outer area consisting of "macrocysts" or packets of cells, which resembled thermophilic *Methanosarcina* and *Methanococcus* strains and in which cell division by formation of one centripetally growing cell wall was observed; B, a loosely packed central region of ovoid cells with intercellular spaces, which could form C, extensive internal cavities, containing non-methanogenic rods; these were the exclusive sites of gas formation.

The pellet surfaces were covered by a fibrillar meshwork, part of which was believed to be bacterial flagella or fimbriae, and other regions exopolysaccharide

Table 6. Biomass forms in anaerobic bioreactors

System of biomass retention	Biomass form	Anaerobic reactor type
Sedimentation	Floc	Contact process
Sedimentation	Pellet/granule	UASB
Adsorption by non-porous inert-support medium	Biofilm	Anaerobic filter RBC AAFEB AFB
Entrapment by porous inert support medium	Biofilm	Anaerobic filter AFB

fibres. Gas bubble formation always originated in the interior of the granule, passing via an orifice to the outer surface, subsequent to release.

The importance of extracellular polymeric material to adhesion and film and floc formation has already been discussed; the aggregation of bacteria into pellets under the low-shear conditions of the UASB reactor system may also be strongly influenced by polymers [50], as microbial agglutination and accumulation of extracellular material have been shown to be related [20].

Pellets, therefore, appear to be granular consortia of various types of bacteria, whose interspecific relationships have not been fully clarified, and whose formation depends critically upon wastestream composition and digester conditions.

The various forms of biomass presently utilised in anaerobic waste treatment are outlined in Table 6.

3.5 Entrapment in Natural Polymers

The immobilisation of cells by entrapment in natural polymers such as polysaccharides has proved successful in industrial processes where cell viability is important [54]. This technique could be extended to systems for the treatment of industrial wastewaters, as the biofilm processes used at present are, in essence, cells entrapped or immobilised in gels that they themselves produce.

Collagen, cellulose acetate, polyacrylamide and calcium alginate have been employed in various systems. Covalent modification is necessary in the entrapment of bacteria in collagen, to stabilise the cell-collagen complex; the stabilising agent most frequently used is gluteraldehyde [55]. Cross-linkage of cells and support is believed to occur via amino groups on the collagen, and the immobilised cells can be rolled into sheets for use in industrial reactors [54].

Cells trapped in cellulose acetate lose some of their activity due to contact with the organic solvent used to emulsify the acetate, but immobilisation has been successfully carried out [56]. The resultant fibres formed contain the cells trapped in aqueous suspension in microscopic cavities within the polymer.

53

Immobilisation in polyacrylamide may also prove damaging to metabolic activity, as the cross-linking agents utilised tend to be extremely reactive and produce concentrations of free radicals. However, the material can be polymerised to give a range of pore sizes, and trapped cells can grow within the complex [54].

Entrapment of biomass in gels of calcium alginate is considered the least disruptive of the polymer immobilisation processes: the cells are suspended in a solution of sodium alginate and the suspension passed into a calcium salt solution [57]. The bead or strand of alginate gels immediately at the surface and the interior of the structure gels within 0.5 h [58]. Calcium alginate is destroyed by chelating agents, presenting a drawback to its use in immobilisation, but strontium or barium, when used as the cross-linking agent, can overcome this in some instances [59]. Alginates are natural copolymers of D-mannuronate and L-guluronate, the ratio of the former to the latter varying considerably from source to source. Generally, alginates suitable for immobilisation tend to be "high guluronic", as only the L-guluronate moieties can gel with calcium ions [54].

The use of natural polymers may have several advantages as support media over other generally utilised materials: their worth has yet to be tested.

3.6 Estimation of Microbial Mass and Activity

The assessment of specific microbial biomass in anaerobic digesters presents two major difficulties; firstly, in some systems, the bacteria are attached to small surfaces (carrier media) and secondly they are generally present as consortia of different physiological and morphological types. The determination of biomass and community composition requires extraction, isolation and separation of biochemical constituents which are specific to particular microbial groups. Cellular components which change rapidly in nature upon cell death can be used, for example, to estimate viable biomass.

Overall biomass may be estimated by the measurement of constituents common to all cells, such as ATP, which is rapidly metabolised in living cells and which can be determined by the sensitive luciferin-luciferase assay [60]. Inactivation of the ATPase enzymes whilst leaving the assay system unaltered is necessary but has been found to be problematical [61]. Acid extraction techniques cannot be utilised for the estimation of biomass growing as films on metallic surfaces; a chemical procedure was therefore devised using chloroacetaldehyde to form a 1-N^6-ethenoadenosine derivative which permitted detection by fluorescence in the 10^{-12} molar range [61]. Separation of the components by high pressure liquid chromatography (HPLC) using fluorescence detectors was then carried out. These assays may be complicated by the presence of extracellular adenosine nucleotides [60].

Microbial membranes contain a relatively constant proportion of phospholipids [62] which are not found in storage lipids, and which have a relatively swift turnover in both viable and non-viable cells [62–64]. Phospholipids comprise 98%

54

of eubacterial membrane lipids and 50% of the lipids of other microorganisms, and can be extracted and quantified. Perchloric acid digestion, followed by colorimetric analysis provides an estimate of lipid phosphate to a sensitivity of 10^{-9} moles [62], which corresponds to the detection of approximately 10^9 bacteria of the size of *E. coli*. Increased sensitivity, to 10^{-11} moles phospholipid (i.e., lipid phosphates in ca. 10^6 bacteria), was obtained by the application of the acid followed by hydrofluoric acid hydrolysis using gas liquid chromatography (GLC) to determine the glycerol [60]. Palmitic acid, a component of virtually all bacterial lipids [65] is amenable to assay by GLC to 10^{-12} moles (5×10^5 bacteria of the size of *E. coli*) [66], and the use of the muramic acid in prokaryotic cell walls as a biomass indicator has been discussed [67, 68, 69].

The Gram-negative bacterial biomass of sediments has been assessed by trichloroacetic acid or phenol water extraction of lipopolysaccharide followed by the assay of ketodeoxyoctonate, β-hydroxymyristate and anti-human complement activity [70]. Cold concentrated hydrofluoric acid has been used to liberate ribitol and glycerol specifically and quantitatively from the lipid-extracted residue of the Gram-positive bacteria producing teichoic acids [60]. This assay was sensitive down to 10^{-11} moles and corresponded to 5×10^6 bacteria similar in dimensions to *Staphylococcus aureus*.

The assessment of the biomass of a particular part of digester microflora may be necessary in some instances to determine the relative degrees of microbial activity predominant under certain conditions. The use of lipid component analysis ("signature lipids") has permitted the separation of prokaryotic from eukaryotic organisms, and estimations of the ratios of aerobic to anaerobic microorganisms [60]. These can be extended to anaerobic systems: anaerobic fermentative bacteria have been found to possess unique phosphosphingolipids [71–73], which can be isolated, and derivative branched sphinganines prepared and subsequently assayed by GLC [74]. The identification of methanogenic bacteria by HPLC assay of extracted, purified and derivatised di- and tetra-phytanyl glycerol ether phospholipids has been cited [60]. Methods of estimation of the proportion of acetoclastic methanogenic bacteria (i.e., those converting acetate to CH_4 and CO_2) in anaerobic digestion systems have been proposed [75, 76]. A simple technique, which comprises the addition of increasing amounts of acetate to a series of sludge samples and the determination of the rate of CH_4 production, takes advantage of the findings that biomass growth is minimal during a maximum 24 h incubation [77] and that the conversation rates of acetate are not affected by certain limited substrate concentrations and obey zero order kinetics [77, 79].

Optimisation of the anaerobic digestion process depends largely on control of the rate-limiting step of the series of reactions comprising degradation; for most waste-types, this step is considered to be methanogenesis (see Chap. 1). The use of fluorescence as a tool in the monitoring of methanogenic activity in digester sludges has been reported [76, 80] and its potential for measurement of active methanogenic biomass assessed [83].

The growth of microbial films around support particles in a fixed-film reactor can be substantial and an estimation of the biomass present in the system is necessary for the proper control of waste processing. Theoretical relationships relating bed height and biofilm volume in a fluidised bed reactor have been reported, which compare well with experimentally estimated values [82]. According to Rittmann and McCarty [83, 84], the rate of substrate utilisation is dependent upon the mass of biofilm present, and biofilm growth must be included in a model if the variation of biofilm mass with time is to be determined. Equations were presented for the calculation of biofilm thickness at steady state substrate concentrations greater than a minimum substrate concentration below which steady state biofilm could not exist.

The above model was subject to criticism because it included only biomass decay as a mechanism of film loss. As other loss mechanisms such as shear stress and sloughing are also of importance in biofilm loss rate and steady-state thickness, the biofilm model was reassessed and shown to be applicable to situations where shear losses were significant [85].

Estimation of biological mass, composition and activity in digester systems are necessary to the optimisation of anaerobic waste processing, thus techniques which improve these estimations will also help to increase the level of control of operational parameters possible in such systems.

References

1. Bungay HR, Bungay ML, Haas CN (1983) Engineering at the microorganism scale. In: Tsao GT (ed) Ann reports on fermentation processes, vol 6. Academic Press, New York London, p 149
2. Rutter PR (1980) The physical chemistry of the adhesion of bacteria and other cells. In: Curtis ASG, Pitts JD (eds) Cell adhesion and motility: British Soc for Cell Biol, Symp 3. Cambridge University Press, Cambridge, p 103
3. Dolowy K (1980) A physical theory of cell-cell and cell-substratum interactions. In: Curtis ASG, Pitts JD (eds) Cell adhesion and motility: British Soc for Cell Biol, Symp 3. Cambridge University Press, Cambridge, p 39
4. Characklis WG, Cooksey KE (1983) Adv Appl Microbiol 29:93
5. Marshall KC (1976) Interpretation in microbiol ecology. Harvard University Press, Cambridge, Mass London, p 1
6. Fletcher M (1979) The attachment of bacteria to surfaces in aquatic environments. In: Ellwood DC, Melling J, Rutter P (eds) Adhesion of microorganisms to surfaces. Academic Press, London New York San Fransisco, p 87
7. Derjaguin BV, Landau L (1941) Acta Physicochimica URSS 14:633
8. Verwey EJW, Overbeek JThG (1948) Theory of the stability of lyophobic colloids. Elsevier, Amsterdam
9. Rogers HJ (1979) Adhesion of microorganisms to surfaces: some general considerations on the role of the envelope. In: Ellwood DC, Melling J, Rutter P (eds) Adhesion of microorganisms to surfaces. Academic Press, London New York San Fransisco, p 29
10. Dexter SC, Sullivan JD, Williams J, Watson SW (1975) Appl Microbiol 30:298
11. Pethica BA (1979) Microbial cell adhesion. In: Berkeley RCW, Lynch JM, Melling J, Rutter PR, Vincent B (eds) Microbial adhesion to surfaces. Ellis Horwood, Chichester, p 19
12. Lips A, Jessup NE (1979) Colloidal aspects of bacterial adhesion. In: Ellwood DC, Melling J, Rutter P (eds) Adhesion of microorganisms to surfaces. Academic Press, London New York San Francisco, p 5
13. Dehneke B (1975) J Colloid Interface Sci 40:1
14. Derjaguin BV, Muller VM, Toporov YP (1975) J Colloid Interface Sci 53:314
15. Krupp H (1967) Adv Colloid Interface Sci 1:111
16. Trulear MG, Characklis WG (1982) J Water Pollut Control Fed 54:1288
17. Costerton JW (1984) Dev Ind Microbiol 25:363

18. Humphrey BA, Dickson MR, Marshall KC (1979) Arch Microbiol 120:231
19. Nishikawa S, Kuriyama M (1968) Water Res 2:811
20. Brown MJ, Lester JN (1979) Water Res 13:817
21. Pavoni JL, Tenney MW, Echelberger WF (1972) J Water Pollut Control Fed 44:414
22. Sutherland IW (1977) Bacterial polysaccharides. In: Sutherland IW (ed) Surface carbohydrates of the prokaryotic cell. Academic Press, New York, p 27
23. Marrie TJ, Nelligan J, Costerton JW (1982) Circulation 66:1339
24. Costerton JW, Irvin RT, Cheng K-J (1981) Ann Rev Microbiol 35:399
25. Ruseska I, Robbins J, Costerton JW, Lashen ES (1982) Oil Gas J 80(10):253
26. Howell JA, Atkinson B (1976) Water Res 18:307
27. Powell MS, Slater NKH (1982) Biotechnol Bioeng 24:2527
28. Messing RA (1983) Bioenergy production and pollution control with immobilized microbes. In: Tsao GT (ed) Ann reports on fermentation processes, vol 6. Academic Press, New York London, p 23
29. Messing RA, Stineman TL (1983) Annals NY Acad Sci 413:501
30. Pike EB, Curds CR (1971) Soc Appl Bacteriol Symp 1:123
31. Melik DH, Fogler HS (1984) J Colloid Interface Sci 101:72
32. Melik DH, Fogler HS (1984) J Colloid Interface Sci 101:84
33. Daniels S (1974) AIChE Symp Ser 70(136):266
34. Ash SG (1979) Adhesion of microorganisms in fermentation processes. In: Ellwood DC, Melling J, Rutter P (eds) Adhesion of microorganisms to surfaces. Academic Press, London, New York San Francisco, p 57
35. Forster CF (1968) Water Res 2:767
36. Forster CF (1971) Water Res 5:861
37. McLoughlin AJ, Vallom JK (1984) J Appl Bacteriol 57:485
38. Mitchell P (1966) Biol Rev 41:445
39. Hamilton WA (1977) Energy coupling in substrate and group translocation. In: Haddock BA, Hamilton WA (eds) Microbial energetics: 27th Symp Soc for Gen Microbiol. Cambridge University Press, Cambridge, p 185
40. Treweek GP, Morgan JJ (1977) J Colloid Interface Sci 60:258
41. Napper DH (1977) J Colloid Interface Sci 58:390
42. Pethica BA (1961) Expt Cell Res Suppl 8:123
43. Tenney MW, Stumm WJ (1965) J Water Pollut Control Fed 32:1370
44. Busch PL, Stumm WJ (1968) Environ Sci Technol 2:49
45. Jewell WJ, Switzenbaum MS, Morris JW (1981) J Water Pollut Control Fed 53:482
46. Parker DS, Kaufmann WJ, Jenkins D (1971) J Water Pollut Control Fed 43:1817
47. Cohen A, Zoetemeyer RJ, van Deursen A, van Andel JG (1979) Water Res 13:571
48. Lettinga G, van Velsen AFM, de Zeeuw W, Hobma SW (1979) Feasibility of the upflow anaerobic sludge blanket (UASB) process. In:Proc 1979 Nat Conf on Environ Eng July 9–11. ASCE, San Francisco, p 35
49. Pipyn P, Verstraete W (1979) Biotechnol Lett 1:495
50. Ross WR (1984) Water SA 10:197
51. Hulsoff-Pol LW, de Zeeuw WJ, Velzeboer CTM, Lettinga G (1983) Water Sci Technol 15:291
52. Klapwijk A, Smit H, Moore A (1981) Denitrification of domestic wastewater in an upflow sludge blanket reactor without carrier material for the biomass. In: Cooper PF, Atkinson B (eds) Biological fluidised bed treatment of water and wastewater. Ellis Horwood, Chichester, p 205
53. Bochem HP, Schoberth SM, Sprey B, Wengher P (1982) Can J Microbiol 28:500
54. Bucke C (1983) Biochem Soc Symp 48:25
55. Venkatasubramanian K, Veith WR (1979) Prog Ind Microbiol 15:61
56. Marconi W, Moriski F (1979) Appl Biochem Bioeng 2:219
57. Kierstan M, Bucke C (1977) Biotechnol Bioeng 19:387
58. Cheetham PSJ, Blunt KW, Bucke C (1979) Biotechnol Bioeng 21:2155
59. Paul F, Vignais PM (1980) Enzyme Microbiol Technol 2:281
60. White DC (1983) Symp Soc Gen Microbiol 34:37
61. Davis WM, White DC (1980) Appl Environ Microbiol 40:539
62. White DC, Davis WM, Nickels JS, King JD, Bobbie RJ (1979) Oecologia 40:51
63. King JD, White DC, Taylor CW (1977) Appl Environ Microbiol 33:1177

64. White DC, Bobbie RJ, Morrison SJ, Oosterhof DK, Taylor CW, Meeter DA (1977) Limnol Ocean-ography 22:1089
65. Kates M (1964) Adv Lipid Res 2:17
66. Bobbie RJ, White DC (1980) Appl Environ Microbiol 39:1212
67. Moriarty DJW (1977) Oecologia 26:317
68. King JD, White DC (1977) Appl Environ Microbiol 33:777
69. Fazio SD, Mayberry WR, White DC (1979) Appl Environ Microbiol 38:349
70. Saddler JN, Wardlaw AC (1980) Antonie van Leeuwenhoek 46:27
71. LaBach JP, White DC (1969) Lipid Res 10:528
72. White DC, Tucker AN (1970) Lipids 5:56
73. Rizza B, Tucker AN, White DC (1970) J Bacteriol 101:84
74. White DC, Tucker AN, Sweeley CC (1969) Biochim Biophys Acta 187:527
75. van den Berg L, Lentz CP, Athey RJ, Rooke EA (1974) Biotechnol Bioeng 16:1459
76. Delafontaine MJ, Naveau HP, Nyns EJ (1979) Biotechnol Lett 1:71
77. Valke D, Verstraete W (1983) J Water Pollut Control Fed 55:1191
78. Lawrence PL (1969) J Water Pollut Control Fed 41:R1
79. van den Berg L, Patel GB, Clark DS, Lentz CP (1976) Can J Microbiol 22:1312
80. Binot RA, Naveau HP, Nyns EJ (1981) Biotechnol Lett 3:623
81. Pause SM, Switzenbaum MS (1984) Biotechnol Lett 6:77
82. Tsezos M, Benedek A (1980) Water Res 14:689
83. Rittmann BE, McCarty PL (1980) Biotechnol Bioeng 22:2343
84. Rittmann BE, McCarty PL (1980) Biotechnol Bioeng 22:2359
85. Rittmann BE (1982) Biotechnol Bioeng 24:501

4 Influence of Environmental Factors

Several environment factors can affect anaerobic digestion, either by enhancing or inhibiting parameters such as specific growth rate, decay rate, gas production, substrate utilisation, start-up and response to changes in input.

4.1 Temperature

Temperature is one of the major influences on all of the above. The mesophilic range (25–45 °C) is generally used in anaerobic biological reactor systems as the number of thermophilic anaerobic species is small [1]. Thermophilic bacteria are typically considered to exist and grow within the range 55–80 °C; their enzyme systems are physiologically stable at these elevated temperatures, a condition attributable to the presence of heat-stable macromolecules [2]. The optimum temperature of growth of anaerobic microorganisms is 35 °C or greater, and although anaerobic digesters have been reported to operate at substantially lower temperatures, such as 20 °C [3], anaerobic growth under these conditions is protracted and difficulties in the start-up of some reactors have been reported [4]. In situations, therefore, where reactor operating temperature is low, start-up will benefit if initiated at approximately 35 °C.

The rate of the anaerobic degradation process, in common with other enzyme-mediated bacterial reactions, is subject to the influence of temperature. At values of less than 25 °C, the digestion rate decreases sharply and conventional anaerobic reactors in operation at ambient temperatures in cooler climates may require detention times of as much as 12 weeks for the treatment of sewage sludges [5].

In the range of 25–40 °C, a gradual increase in degradation rate typically occurs, although optimal rates for various microbial species will vary. At temperatures in excess of 45 °C, a rapid decline in digestion rate ensues as the mesophilic growth limit is attained. The active range of thermophilic digester populations under anaerobic conditions has been reported as 50–60 °C, although growth optima are ill-defined [5]. Conditions for maintenance of a stabilised thermophilic fixed-film reactor operation were reported to be improved by careful management of the microorganisms and stringent control of operational parameters [6]; at 55 °C, the expanded bed assembly achieved high solids concentrations and film depths, and moderate (1.5–3.0 g COD 1^{-1}) and high (5.0–16.0 g COD 1^{-1})

strength wastes were treated with maximum removal efficiencies of 80% and 73% respectively.

Mesophilic systems and cultures of rumen microorganisms have been used as sources of thermophilic populations of bacteria, but the mechanisms of metabolic adjustment to the elevated temperature regime were unclear [5]. Selection processes or adaptation of enzyme systems may have occurred. The majority of industrial digester systems operate in the mesophilic range of 30–40 °C. It is probable that increases in microbial reaction rates at the elevated temperatures of thermophilic processes, and hence decreases in SRT, may prove advantageous under some circumstances, although lack of stability in thermophilic municipal waste treatment has been reported [6]. The excess energy necessary for the maintenance of a thermophilic digester in temperate climates, where the system must exist at a temperature some 30 °C greater than ambient, is one of its major drawbacks. Thermophilic digestion is thus most practical where the wastestream to be treated is discharged at an extremely high temperature and the digester is present on site.

The significance of temperature to the rate of anaerobic digestion dictates that the final operating reactor temperature be considered as one of the principle design parameters. In psychrophilic, mesophilic or thermophilic ranges, uniformity of temperature over the entire vessel contents is paramount to anaerobic digestion processes. Rapid alterations in reactor temperature of even a few degrees can result in a marked upset in microbial metabolism and necessitate several days recovery. A consistent temperature throughout the system can be provided by adequate mixing of digester contents by paddle, gas-sparging, or flow over heat exchangers; if localised areas within the vessel deviate from the normal, microbiological activities in these regions will vary and transient pockets of increased temperature may inhibit or inactivate the bacterial population. The inefficiency of climatically-heated small anaerobic digesters may frequently be the result of diurnal fluctuations in temperature [5]. The stabilisation of internal temperatures in large industrial reactor systems in tropical climates may require the incorporation of some form of heating control if effluent quality and gas production are to be uniform. An agitation mechanism, sufficient to disperse any localised buildup of heat in reactor contents is also a necessity. Strong sunshine can cause variations in surface and internal temperature if directed at one side of the reactor vessel.

Seasonal changes in temperature will be accommodated with little effect on the bacterial population if gradual alterations of approximately $1 \,°C\,d^{-1}$ in internal system temperature are introduced; such changes permit microbial adaptation to the novel temperature.

In temperature-controlled anaerobic systems, no alteration in excess of 1 °C should occur as vessel contents pass over or through heating elements. The heat generated by bacterial activity under anaerobic digester conditions is, in theory, sufficient to maintain the environmental temperature in the mesophilic range, but the energy evolved as heat in the conversion of typical wastestreams is inadequate to raise the temperature of a cold influent and also to compensate for heat losses at cool ambient temperatures.

Arrhenius' Law describes the effect of temperature on chemical reaction rates, but may not be an adequate representation of the rate of bacterial growth in relation to temperature. An equation has been suggested relating microbial growth rate and culture temperature (T, in K), of the form:

$$\mu = 6\,(T - T_0), \tag{23}$$

where T_0 is a temperature intrinsically characteristic of bacteria and dependent upon whether they are psychrophiles, mesophiles or thermophiles [7]. The function itself does not set an upper boundary on microbial growth; the authors suggested that it was applicable between a minimum and an optimum growth temperature. Although mesophilic microorganisms may have an upper growth limit of the order of 45 °C, the limit is not strictly defined, and slow growth of thermophilic species at temperature values below 45–50 °C may be expected. An investigation into the laboratory-scale digestion of beef-cattle wastes at temperatures from 30–65 °C indicated that 60 °C was the optimum temperature for thermophilic growth [8]. With the exception of the longest retention time of 18 days, a fall in degradation rates between 60 °C and 65 °C was apparent. At a retention time of 2–5 days, a steady state system with low volatile solids degradation could only be obtained at 60 °C. According to Cooney [9], three statements can be made with regard to the effect of temperature on microbial growth rate. Firstly, growth occurs only over a temperature range that is approximately 15 °C at either side of optimum for any microorganism; secondly, an elevation of temperature occasions a slow increase in rate of bacterial growth until the maximum rate is attained; and thirdly, above this optimal temperature growth rate falls rapidly upon additional temperature increase. This latter is the result of an increased rate of microbial death; the death rate within a microbial population increases much more swiftly with temperature than the rate of growth, and hence overall growth rates rapidly decline above the maximum temperature.

The efficiency of the conversion of carbon-energy substrate to cell mass is also subject to the influence of temperature [9]. The maximum conversion yield, based on investigation into the growth of a yeast and a mixed population of thermophilic bacteria with methanol as substrate, was found to occur at a temperature below that of the maximum growth rate [10]. A number of microbial metabolic systems are also susceptible to the effect of temperature. Cellular components such as ribonucleic acid (RNA) are strongly influenced [11]; the RNA content of the microbial cell relates in a greater degree to the relative growth rate (μ/μ_{max}) than to the actual rate of cell multiplication [12]. This reflects the influence of temperature upon the rate of protein synthesis and the need for maintenance of protein synthesis at a rate that will sustain a given rate of growth [9].

The sensitivity of methanogenic bacteria to alterations in temperature, compared with other microorganisms, is well known, and the anaerobic digestion process is generally considered to be unsuitable for wastewaters which are dilute and at a low temperature [13]. It was reported that careful consideration of SRT would overcome this problem in the anaerobic treatment process: bacterial generation time (SRT^{lim}) was stated to be a function both of bacterial growth yield (Y) and maximum substrate utilisation ($r_{x,\,max}$) and was described by the following equation:

$$SRT^{lim} = (Yr_{x,\,max} - b)^{-1}, \tag{24}$$

61

where b = bacterial decay rate [13]. The growth yield in anaerobic systems is much less than that found in aerobic processes, and $r_{x, max}$ is dependent upon temperature. The doubling time of methanogens may be as long as 30 days at lower temperatures and extended SRTs were thus deemed necessary for the treatment of cold wastewaters which were also dilute.

Several investigators [3, 14] have analysed the low temperature treatment of wastes of dilute (200–600 mg l^{-1}) organic concentrations, and digestion processes were observed to be stable at temperatures around 20 °C. These investigations were accomplished using fixed-film processes of which extended SRTs and short HRTs are one of the major advantages. However, as the majority of industries generate quantities of waste heat as a component of their effluent streams, the maintenance of anaerobic digester temperature above 25 °C presents little difficulty and mesophilic systems can be designed to take advantage of the waste energy.

4.2 Hydrogen Ion Concentration (pH)

pH is an important parameter in microbial metabolism; it is a measure of the activity of hydrogen ions and the majority of bacteria (with the notable exception of most methanogens) are able to grow over a range of three pH units. This represents a hydrogen ion concentration range of 1000–10 000 fold. Most microorganisms also exhibit a pH value at which growth is maximal. This is generally between pH 6.5 and 7.5. At values below 5.0 and in excess of 8.5, growth is frequently inhibited, although some microorganisms, such as a small number of *Acetobacter* spp., are exceptions [9]. A relationship appears to exist between pH and temperature: as the growth temperature of bacteria increases, the observed pH optimum also increases [15]. The control of pH is fundamental to the maintenance of optimal bacterial growth and/or conversion processes in anaerobic microbial systems.

Deterioration of the anaerobic digestion process at pH values of less than 6.5 and greater than 8.2 has been reported [16]. The conditions prevalent at low pH values in anaerobic systems can be highly detrimental to methanogenic populations, and inhibition begins at around pH 6.5 in conventional digesters such as the CSTR. A value of pH 5.5 was reported for the effects of inhibition of CH_4 formation to become apparent in an anaerobic filter system [17], whilst in another investigation, also utilising an anaerobic filter, the optimal pH for the conversion of methanol was found to be 5.5–6.0, with reduction in pH to 4.6 causing a 50% loss of CH_4 formation and a sharp decline in VFA production [18]. The VFA levels remained low, even subsequent to pH adjustment, and system recovery, primarily of acetate formation, was apparent only after 18 days. An increased level of bicarbonate in the system resulted in a measurable increase in VFA concentration only after 5 days, indicating the severe effect of a low pH (i.e. below 4.0) on VFA formation. pH is therefore a major factor governing anaerobic degradation. Two mechanisms of inhibition at low and high pH levels, however, were observed in an investigation with anaerobic filters; the absence of a lag period prior to system recovery was observed after pH restoration from high (basic) pH values, whereas considerable time was required when a low pH period exceeded 3 days [16].

Severe, transient pH shocks of 8 h duration were applied to an anaerobic fluidised bed reactor [19]. Influent pH values were decreased to 3.0, and increased to 10.0 during the course of the investigation; the reactor was reported to remain stable, with an effluent pH increase of 0.18 pH units and an alkalinity increase of 5% recorded. The VFAs increased by 400%, but methanogenesis was not detectably inhibited. The high recycle ratio and the inherent buffering capacity of the fluidised bed configuration promotes increased tolerance of influent fluctuations, especially those of such potentially damaging consequence as pH.

The acidity or alkalinity of the anaerobic reactor contents is the result of acid-base system interactions. These systems can be weak or strong and the acidic and basic components may be present in the influent wastestream, or may be the result of reactions occurring throughout the degradation process [20]. In most anaerobic reactor units, the acid-base systems in operation are the weak carbonic, hydrosulphuric and orthophosphoric. VFAs and ammonia are also included here [21]. The extremely similar dissociation constants of acetic and propionic acids, the main VFAs present, allow them to be considered as one weak acid [20]. The hydrosulphuric and orthophosphoric systems under anaerobic conditions were reported to provide very limited buffering capacity as they existed in extremely low concentrations; between the pH values of 6.0–7.7, negligible buffering functions were provided by ammonia and the VFAs. System buffering under these conditions was almost exclusively the result of the dissociation of carbonic acid [20]. The almost total dissociation of ammonia and VFAs between pH values of 6.0–7.5 enables them to function as strong base and acid systems respectively. The above investigations indicated that the pH under anaerobic conditions was controlled by the association of the carbonic acid system and a net strong base, the latter being the net result of the activities of the VFAs, ammonia and any other strong acids and bases present.

The methanogens have a limited pH range around neutrality, while the hydrolysing organisms exhibit little activity below pH 6.5 in conventional digesters [5]. The acid formers are reported to be intolerant of low pH values, as only 14% of strains isolated from an anaerobic digester in one investigation produced sufficient acid to reduce the pH of an unbuffered medium below pH 6.0; a pH range tolerance of 6.2–9.1 was recorded for *Peptostreptococcus* and *Bacteroides* isolates [22]. In addition, propionate, a VFA intermediate, is toxic to those organisms assimilating H_2.

VFA formation from methanol was observed to be more sensitive to pH shocks in the acidic pH region of 3.5–4.0, supporting the observation of the acid-intolerance of the acid-producing bacteria, as VFA formation recovered from pH shock with more difficulty than CH_4 formation [18]. The initiation of a methanogenic reactor in a two phase assembly should ideally take place at pH 7.0 therefore, in order to encourage maximum methanogen activity. In a separated acidogenic reactor, however, lower pH values are possible, where these may be advantageous to the associated microflora.

4.3 Physical Parameters

In addition to the physical forces that are gravitational, hydrodynamic and electrostatic in nature and intrinsic to a liquid/particle environment, externally applied phenomena also operate. The latter include the energies applied to mix, expand, fluidise, heat and recycle anaerobic reactor vessel contents.

The efficient application of mixing, by fluid flow, gas sparging, paddles or impellers utilises the reactor volume available and minimises the inhibitory effects of local build-up of VFAs and other digestion products. There are several reasons for ensuring the adequate mixing of reactor contents; these include the even dispersion of suspended materials such as granules or flocs to promote maximum biomass/wastewater contact; the rapid and uniform distribution of fresh influent throughout the reactor volume; and also the maintenance of a constant temperature over the contents of the vessel, thus reducing convection currents and thermophoresis. If both influent feedstock and actively metabolising biomass are uniformly dispersed, maximum utilisation of the available substrate is possible and blockage of settling units and effluent pipes is minimised. These problems are most frequently encountered in CSTR, contact and UASB systems (see Chap. 6) where mixing also partly controls the size of flocculated or granulated bioactive particles. Inactive debris can also accumulate at the bottom of these types of reactor if insufficient agitation is applied, and build-up of inorganic solids can decrease effective system volume [5].

The shear forces produced by the external application of energy, in addition to the hydrodynamic shear generated within the digester environment, may be significant. In the sludge blanket configuration, the average concentration of sludge is affected to a small degree by the linear upward fluid velocity. Sludge settling velocity is necessarily much higher than the linear fluid rate and the latter therefore has little significant effect on the upward transport of sludge. However, upward fluid velocity is a function of gas production, being proportionate to the gas formed within the reactor bed: liquid rises in the wake of gas bubbles. Methane and other gases are the final major products of the anaerobic digestion of organic biodegradable materials and in an efficiently operating sludge blanket assembly, the gas produced may provide adequate mixing of the medium. It may also cause turbulence in the settler apparatus that comprises the top part of the reactor, however, and this can lead to biomass washout if the design is not efficient. Mechanical agitation by means of a paddle assembly in a sludge blanket may enhance performance, but the continuous shear so generated may reduce the pellet size.

In a reactor containing flocculated biomass, such as a contact or sludge blanket system, a non-linear relationship is reported to exist between shear stress and shear rate, in low shear situations [23]. The Elastic Floc Model states that the critical shear rate is the point below which the shear field is too weak to force apart two flocculant masses which have collided. Above this critical rate, individual flocs comprise the system; their radii depend upon shear rate but their other characteristics are independent of this parameter. As the shear rate becomes sub-critical, the individual flocs enlarge and become more structured, but their integrity can be assumed to be fixed and they retain their individual status within a group. Thus at low shear rates the large flocs maintain their basic structure, but tend to aggregate [23]. The model is applicable to colloidal particles with a radius of less than 1 μm. The microorganisms in anaerobic digestion are invariably larger than this: the shear stresses encountered at higher shear rates are therefore generally large enough to cause complete floc disintegration.

Efficiency of operation in fixed-film reactors such as the anaerobic filter and expanded and fluidised beds depends greatly on good hydraulic distribution and maximal substrate availability to the digester microorganisms. The minimum linear velocity of the system is that velocity which permits these conditions. As the flow rate increases above the minimum linear velocity, biofilm accumulates on the carrier medium. However, as fluid velocity increases at higher flow rates, biofilm accumulation decreases and detachment can occur, a condition which eventually culminates in process failure. Detachment is also a function of the increase in mass of attached biofilm and has been demonstrated to be dependent upon shear stresses at the biofilm/fluid interface [24].

It has also been recognised that the maximum volume of biofilm attached to the media surface in a turbulent flow regime could be limited by fluid shear stress: the increase of shear removal rate was found to be proportional to interfacial fluid shear stress and, at high film volumes, detachment approached infinity [24, 25]. The critical shear stress in fixed biofilm systems is that value at which all the cells are removed; the proportion of the initial bacterial population remaining after a specified period of shear is indicated by the adhesion number [26]. The provision of settling time in a reactor system prior to operational processes tends to increase the level of shear stress necessary for film removal from media surfaces as the microflora have formed stable linkages. Gram-positive bacteria are more easily removed than those with Gram-negative cell walls in complex media, as the outer surfaces of the latter probably enhance adhesion [27]. Frictional resistance to flow is characteristic of biofilm-dependent reactor processes. This resistance has been observed to increase where filamentous forms of bacteria predominate, possibly as a result of the dissipation of energy by the fibrillar structure [24].

During the accumulation process at the initiation of digester operation, and as the reactor proceeds towards steady state, a succession of microbial communities occurs. The fluid shear stress in the system may enhance this process by shearing and sloughing off microorganisms displaced by other competitive groups. The decay of bacteria within the film also affects biomass loss rate; the steady-state biofilm model of Rittmann and McCarty [28, 29] has been shown to apply also to situations of significant shear-stress losses [30]. The model predicts the conditions under which shear-stress loss is likely to be important.

The physical forces resulting from reactor operation vary under different conditions and with different digester configurations. These forces are the consequence primarily of externally applied energies and, as such, can be controlled by proper attention to operational parameters and careful system design.

4.4 Nutrients

In addition to fundamental requirements for macronutrients such as carbon and nitrogen, the inability of a great number of anaerobes to synthesise some essential vitamin or amino acid necessitates the supplementation of the bacterial medium with certain specific nutrients for growth and activity. The minimum level of any necessary nutrient, which is that concentration which will support a desired growth rate, must be maintained although a maximum level of nutrient requirement also exists above which substrate inhibition occurs. In anaerobic reactor systems, the definitions of these boundaries are nebulous or non-existent. The estimation of mineral requirements is particularly difficult due to the occurrence of cellular and noncellular metal complexes [9]. Phosphates and sulphates can also exert catabolite repression of specific metabolic pathways, while not affecting growth.

Microbial nutritional requirements may also be interdependent: in the presence of low concentrations of potassium, for example, sodium may be able to satisfy some of the potassium requirement. In addition, nutrients not obligatory for growth, such as calcium, may nonetheless be needed for process stability. Substantial excess of nutrients frequently occurs in industrial wastewaters, the prime example being phosphate, but others, such as zinc and copper ions may often be near limiting values.

The identification of specific nutrient requirements can be a time-consuming process and little has been reported relative to anaerobic digester systems. The stimulation of acetate conversion to CH_4 and CO_2 in a mixed population of anaerobic fixed-film digester bacteria was investigated [31]. The addition of nickel, cobalt and molybdenum increased total gas (including CH_4) production by 42%, and allowed greater volumes of food processing waste to be effectively treated by decreasing the reactor residence time. These effects were considered the result of the accumulation of a thicker methanogenic fixed-film. The addition of nickel to a culture of methanogens was reported to increase the level of acetate utilisation. Where this stimulatory effect was observed, *Methanosarcina* spp. were dominant [32]. Nitrate reduction and COD utilisation increased significantly when molybdenum and selenium were added to an anaerobic system [33] and sulphate reducers were observed to require biotin and yeast extract for function [34].

The presence of trace metals such as molybdenum, selenium, tungsten and nickel is probably necessary for the activity of several enzyme systems. The facultative anaerobe *Escherichia coli* is capable of obtaining energy for growth from electron-transport-dependent ATP synthesis under anaerobic conditions using either fumarate or nitrate as a terminal electron acceptor [35]. The two anaerobic electron transport chains utilising nitrate and fumarate appear to share common components. Under anaerobic conditions of molybdenum-limitation, the synthesis of the enzymes formate dehydrogenase and nitrate reductase by *E. coli* is decreased, as both of these enzymes need molybdenum as a cofactor for activity. The dehydrogenase also requires selenium and limitation of this element under anaerobiosis results in a decrease in the enzyme activity [36].

66

Table 7. Some trace element-dependent enzymes of anaerobic bacteria

Element	Enzyme	Anaerobic microorganism	Ref.
Selenium	Formate dehydrogenase	Acetogenic bacteria	Wagner and Andreesen [39]
			Ljungdahl and Andreesen [40]
		Methanococcus vanielii	Jones and Stadtman [41]
	Glycine reductase	Several clostridia	Turner and Stadtman [42]
			Stadtman [43]
			Dürre and Andreesen [44]
	Hydrogenase	*Methanococcus vanielii*	Yamazaki [45]
	Nicotinic acid hydroxylase	*Cl. barkeri*	Imhoff and Andreesen [46]
	Xanthine dehydrogenase	Some clostridia	Wagner and Andreesen [47]
			Dürre et al. [48]
			Dürre and Andreesen [49]
Tungsten	Formate dehydrogenase	Acetogenic bacteria	Andreesen and Ljungdahl [50]
			Andreesen et al. [51]
			Leonhardt and Andreesen [52]
		Methanococcus spp.	Jones and Stadtman [53]
		Cl. thermoaceticum	Ljungdahl and Andreesen [40, 54]
			Yamamoto et al. [55]
Nickel	Carbon monoxide dehydrogenase	*Cl. pasteurianum*	Diekert et al. [56]
			Drake [57]
		Cl. thermoaceticum	Drake et al. [58]
			Diekert and Ritter [59]
		Acetobacterium woodii	Diekert and Ritter [60]
	Hydrogenase	*Methanobacterium thermoautotrophicum*	Graf and Thauer [61]
			Albracht [62]
			Jacobson et al. [63]
		Vibrio succinogenes	Unden et al. [64]
		Desulfovibrio gigas	Cammack et al. [65]
		D. desulfuricans	Krüger et al. [66]
	Methyl-CoM methyl reductase	Methanogenic bacteria	Diekert et al. [67]
			Whitman and Wolfe [68]
			Ellefson et al. [69]

Tungsten acts as a competitive inhibitor of molybdenum in many biological systems [37]. The anaerobic growth of *E. coli* in a medium containing tungstate resulted in the synthesis of inactive formate dehydrogenase and nitrate reductase enzymes. Subsequent incubation of the *E. coli* cells with molybdenum under non-growing conditions led to the rapid activation of both the enzyme systems. Investigations into the requirements of anaerobic bacteria for trace elements have resulted in the discovery of several selenium-, tungsten- and nickel-enzymes; these are tabulated above (Table 7). Both selenium and nickel are often found as contaminants in culture media, the former being present in significant quantities in commercially-produced sodium- and hydrogen-sulphide and the latter in iron salts [38].

Iron limitation has been reported to decrease the levels of hydrogenase, and consequently formate hydrogen lyase, in *E. coli* under anaerobic conditions [35].

Iron was found, however, to be unnecessary for the stimulation of the sulphate-reducing bacteria in an investigation into the nutrient requirements of these organisms, although casamino acids were essential [34]. A VFA mixture (acetate, propionate, n-butyrate, n-valerate, isovalerate, isobutyrate, and 2-methylbutyrate) was not required for growth and was found to be slightly inhibitory. Haemin was also unnecessary, although biotin was essential for one *Desulfovibrio* strain and *p*-aminobenzoic acid (PABA) was stimulatory. Vitamins, VFAs and haemin act as growth factors for most other digester microorganisms, but the bacteria comprising the biofilm in a fixed-film system may slough their extracellular polymeric material in response to shock doses [25]. It was reported that upon heavy increase of growth substrate, biofilm material immediately detached, although cell numbers remained constant and specific substrate removal rate and product formation rate increased instantaneously [70]. The formation of VFAs when no trace elements are present in the digester is negligible. A significant increase in production of VFAs occurred in one experiment after changing to a freshly-prepared solution of essential elements [71]; this provided a strong indication of the importance of these nutrients to the efficient functioning of a stable digestion system.

If the waste to be treated by anaerobic degradation is not rich in micronutrients, supplementation, either chemically or by the addition of another wastestream [72] may be necessary. The design and optimisation of the growth environment, however, should include consideration of the technical and economic constraints imposed by the nature and quantity of the waste to be converted.

References

1. Zinder SH, Anguish T, Cardwell SC (1984) Appl Environ Microbiol 47:808
2. Zeikus JG (1977) Microbiol Rev 41:514
3. Switzenbaum MS, Jewell WJ (1980) J Water Pollut Control Fed 52:1953
4. Salkinoja-Salonen MS (1982) The behaviour of chloroorganic chemicals in the environment and their biological treatment, presented at: FAST environmental biotechnology workshop, Versailles
5. Hobson PN (1982) Production of biogas from agricultural wastes. In: Subba Rao NS (ed) Advances in agricultural microbiology. Butterworth Scientific, London, p 523
6. Schraa G, Jewell WJ (1984) J Water Pollut Control Fed 5:226
7. Rathowski DA, Olley J, McMeekin TA, Ball AJ (1982) J Bacteriol 149:1
8. Varel VH, Hashimoto AG, Chen YR (1980) Appl Environ Microbiol 40:217
9. Cooney CL (1981) Growth of microorganisms. In: Rehm H-J, Reed G (eds) Biotechnology, a comprehensive treatise in 8 volumes, 1: microbial fundamentals. Verlag Chemie, Weinheim-Deerfield Beach, Florida Basel, p 73
10. Snedecor B, Cooney CL (1974) Appl Microbiol 27:1112
11. Tempest DW, Hunter JR (1965) J Gen Microbiol 41:267
12. Tempest DW (1976) The concept of relative growth rate: its theoretical basis and practical application. In: Dean ACR, Ellwood DC, Evans CGT, Melling J (eds) Continuous culture 6: applications and new fields. Ellis Horwood, Chichester, p 349
13. Parkin GF, Speece RE (1983) Water Sci Technol 15:261
14. Jewell WJ, Switzenbaum MS, Morris JW (1981) J Water Pollut Control Fed 53:482
15. Brock TD (1967) Science 158:1012
16. Duarte AC, Anderson GK (1983) I Chem Eng Symp 77:149

17. Clarke RH, Speece RE (1971) The pH tolerance of anaerobic digestion. In: Proc 5th Int Wat Pollut Res Conf. Pergamon, New York, p II/27
18. Lettinga G, van der Geest ATh, Hobma S, van der Laan J (1979) Water Res 13:725
19. Bull MA, Sterritt RM, Lester JN (1983) J Chem Technol Biotechnol 33B:221
20. Capri MG, Marais CvR (1975) Water Res 9:307
21. Pohland FG (1969) High rate digestion control techniques for evaluating acid-base equilibrium. In: Proc 24th Ind Waste Conf Purdue Univ., Lafayette, Indiana 1969. Ann Arbor Science, Ann Arbor Michigan, p 353
22. Iannotti EL, Fischer JR (1984) Effects of ammonia, volatile acids, pH and sodium on growth of bacteria isolated from a swine manure digester. In: Proc 40th Gen Meeting of Soc Ind Microbiol, Sarasota, Florida, Aug. 1983. Victor Graphics, Baltimore, p 741
23. Ekdawi N, Hunter RJ (1983) J Colloid Interface Sci 94:355
24. Trulear MG, Characklis WG (1982) J Water Pollut Control Fed 54:1288
25. Characklis WG, Cooksey KE (1983) Adv Appl Microbiol 29:93
26. Powell MS, Slater NKH (1982) Biotechnol Bioeng 24:2527
27. Rogers HJ (1979) Adhesion of microorganisms to surfaces: some general considerations on the role of the envelope. In: Ellwood DC, Melling J, Rutter P (eds) Adhesion of microorganisms to surfaces. Academic Press, London New York San Francisco, p 29
28. Rittmann BE, McCarty PL (1980) Biotechnol Bioeng 22:2343
29. Rittmann BE, McCarty PL (1980) Biotechnol Bioeng 22:2359
30. Rittmann BE (1982) Biotechnol Bioeng 24:501
31. Murray WD, van den Berg L (1981) Appl Environ Microbiol 42:502
32. Speece RE, Parkin GF, Gallagher D (1983) Water Res 17:677
33. Chakrabarti T, Jones PH (1983) Water Res 17:931
34. Ueki A, Suto T (1981) J Gen Appl Microbiol 27:229
35. Haddock BA (1977) The isolation of phenotypic and genotypic variants for the functional characterisation of bacterial oxidative phosphorylation. In: Haddock BA, Hamilton WA (eds) 27th Symp Soc Gen Microbiol, London 1977. Cambridge University Press, Cambridge, p 95
36. Lester RL, De Moss JA (1971) J Bacteriol 105:1006
37. Scott RH, De Moss JA (1976) J Bacteriol 126:478
38. Thauer RK, Morris JG (1984) Metabolism of chemotrophic anaerobes: old views and new perspectives. In: Kelly DP, Carr NG (eds) 36th Symp Part II Soc Gen Microbiol, Warwick 1984. Cambridge University Press, Cambridge, p 123
39. Wagner R, Andreesen JR (1977) Arch Microbiol 114:219
40. Ljungdahl LG, Andreesen JR (1978) Formate dehydrogenase, a selenium-tungsten enzyme from *Clostridium thermoaceticum*. In: Fleischer S, Parker L (eds) Methods in enzymology 53. Academic Press, New York London, p 360
41. Jones JB, Stadtman TC (1981) J Biol Chem 256:656
42. Turner DC, Stadtman TC (1973) Arch Biochem Biophys 154:366
43. Stadtman TC (1978) Selenium-dependent clostridial glycine reductase. In: Fleischer S, Packer L (eds) Methods in enzymology 53. Academic Press, New York London, p 373
44. Dürre P, Andreesen JR (1982) J Gen Microbiol 128:1457
45. Yamazaki S (1982) J Biol Chem 257:7926
46. Imhoff D, Andreesen JR (1979) FEMS Microbiol Lett 5:155
47. Wagner R, Andreesen JR (1979) Arch Microbiol 121:255
48. Dürre P, Andersch W, Andreesen JR (1981) Int J Systematic Bacteriol 31:184
49. Dürre P, Andreesen JR (1982) Arch Microbiol 131:255
50. Andreesen JR, Ljungdahl LG (1973) J Bacteriol 116:867
51. Andreesen JR, El Ghazzawi E, Gottschalk G (1974) Arch Microbiol 96:103
52. Leonhardt U, Andreesen JR (1977) Arch Microbiol 115:277
53. Jones JB, Stadtman TC (1976) *Methanococcus vanielli*: growth and metabolism of formate. In: Schlegel HG, Gottschalk G, Pfennig N (eds) Symp on microbial production and utilization of gases (H_2, CH_4, CO). p 199
54. Ljungdahl L, Andreesen JR (1975) FEBS Lett 54:279
55. Yamamoto I, Saiki T, Liu S-M, Ljungdahl LG (1983) J Biol Chem 258:1826
56. Diekert GB, Graf EG, Thauer RK (1979) Arch Microbiol 122:117
57. Drake HL (1982) J Bacteriol 149:561

58. Drake HL, Hu S-I, Wood HG (1980) J Biol Chem 225:7174
59. Diekert G, Ritter M (1983) FEBS Lett 151:41
60. Diekert GB, Ritter M (1982) J Bacteriol 151:1043
61. Graf EG, Thauer RK (1981) FEBS Lett 136:165
62. Albracht SPJ, Graf EG, Thauer RK (1982) FEBS Lett 140:311
63. Jacobson FS, Daniels L, Fox JA, Walsh CT, Orme-Johnson WH (1982) J Biol Chem 257:3385
64. Unden G, Böcher R, Knecht J, Kröger A (1982) FEBS Lett 145:230
65. Cammack R, Patil D, Aguirre R, Hatchikian EC (1982) FEBS Lett 142:289
66. Krüger H-J, Huynh BH, Ljungdahl PO, Xavier AV, De Vartanian DV, Moura I, Peck HD Jr, Teix- eira M, Moura JJG, Le Gall J (1982) J Biol Chem 257:1462
67. Diekert G, Klee B, Thauer RK (1980) Arch Microbiol 124:103
68. Whitman WB, Wolfe RS (1980) Biochem Biophys Res Comm 92:1196
69. Ellefson WL, Whitman WB, Wolfe RS (1982) Proc Nat Acad Sci USA 79:3707
70. Bakke R (1983) MS Thesis, Montana State University, Bozeman, Montana, USA
71. Lettinga G, de Zeeuw W, Ouborg E (1981) Water Res 15:171
72. Byrd JF (1961) Combined treatment. In: Proc 16th Ind Waste Conf, Purdue Univ, Lafayette, In- diana. Ann Arbor Science, Ann Arbor Michigan, p 92

5 Toxic Substances in Anaerobic Digestion

Inhibition of the anaerobic digestion process can be mediated to varying degrees by toxic materials present in the system; these substances may be components of the influent wastestream, or byproducts of the metabolic activities of the digester bacteria. Inhibitory toxic compounds include sulphides, consequential in the processing of wastes from such sources as molasses fermentation, petroleum refining and tanning industries, and volatile acids, microbial products which can accumulate and exceed the reactor buffering capacity. Inhibition may also arise as the consequence of the levels of ammonia, alkali and the alkaline earth metals, and heavy metals in the system. The latter have been considered the most common and major factors governing reactor failure [1, 2].

The toxic content of wastestreams, which may include substances such as pentachlorophenol [3] and hemicellulose derivatives [4] principally obstructs the activities of the sensitive OHPA and methanogenic portions of the digester population. A relatively short-term test has, however, been evaluated to assess the toxicity of a specific organic compound to an active, although unacclimated microflora in sludge of municipal origin [5]; data presented have indicated acclimation to, or recovery from, applications of such compounds as formaldehyde and chloroform [6]. The need for development of adequate mathematical models for design and operation of biological processes for the refining of wastewaters containing toxic compounds has been emphasised [7], and the inhibition of the initial enzymatic phases of anaerobic conversion by hydrolysis products such as alkanols [8] requires more investigation.

5.1 Volatile Acids Inhibition

Anaerobic reactor instability is generally manifested by a marked and rapid increase in VFA concentrations; this is frequently indicative of the failure of the methanogenic population due to other environmental disruptions such as shock loadings, nutrient depletion or infiltration of inhibitory substances. Acetate has been described as the least toxic of the volatile acids [9], whilst propionate has often been implicated as a major effector of digester failure [10, 11]. Microbial growth was observed to be significantly inhibited at 35 g l^{-1} acetate in one investigation [9] although sudden concentration increases in either acetic or n-butyric acid reportedly caused stimulation rather than inhibition of methanogens in another [10]. Propionate was found to be more inhibitory than butyrate for *Bacteroides*, but the reverse applied in the case of *Peptostreptococcus* [9]. Methanogenic populations were demonstrated to be inhibited at propionate concentrations in excess of 3000 mg l^{-1}, although this effect could be overcome by acclimation [10]. The methanogen *Methanobacterium formicicum* was reported to tolerate both acetate and butyrate at concentrations of up to 10000 mg l^{-1} [11] although variable inhibitory levels for propionate of 1000 mg l^{-1} [9] and 5000 mg l^{-1} [11] have been recorded.

In any system that is overloaded and hence contains high concentrations of VFAs, the digestion process may be inhibited by the VFAs themselves and also by any alcohols remaining undegraded. A low pH is more detrimental to the degradation of the common VFAs than it is to CH_4 formation from methanol, as the latter does not dissociate within the pH range of 3.0 to 8.0. According to Andrews [12], conversion-rate inhibition by the volatile acids at acidic pH values can be attributed to the existence of unionised VFAs in significant quantities in the system. These unionised acids are present in amounts dependent upon the total concentration of VFAs in solution. The undissociated nature of these acids allow them to penetrate the bacterial cell membrane more efficiently than their ionised counterparts, and once assimilated, induce an intracellular decrease in pH and hence a decrease in microbial metabolic rate.

A pH-dependent equilibrium exists between the ionised and unionised components of the volatile acids:

$$CH_3COOH \rightleftharpoons CH_3COO^- + H^+. \tag{25}$$

As the pH value drops, equilibrium shifts to the left, resulting in an increase in the concentration of unionised volatile acids (UVA). A value of 1.73×10^{-5} has been given for the ionisation constant at 35 °C. Digester failure becomes increasingly more likely as the concentration of unionised acids rises above 10 mg l^{-1} [13]. These results are indicative of a more direct functional relationship between the concentrations of UVAs and the level of toxicity (defined as a total cessation of microbial metabolic activity) within the digester system.

5.2 Sulphide Inhibition

The sulphates and other oxidised compounds of sulphur are easily reduced to sulphide under the conditions prevalent in anaerobic digesters. Methionine and cysteine, the sulphur-containing amino acids of protein, can also undergo degradation to sulphide. These compounds are of significance when anaerobic treatment is considered for industrial processes which tend to produce large quantities of sulphides in their wastestreams (see Chap. 11). The sulphides formed by the activity of reactor microorganisms may be soluble or insoluble, depending upon their associated cations. When the salts formed are insoluble, as are many of the metal sulphides, they have negligible effects on digestion. Iron addition, for example, can ameliorate sulphide inhibition by removing the S^{2-} ion from solution.

Desulfovibrio and other sulphate-reducing genera form sulphides from sulphates and some of the fermentative microorganisms utilise the sulphur-containing amino acids to produce sulphides. In completely mixed and anaerobic filter reactors, sulphide concentrations of 1500 mg l^{-1} proved to be toxic to the reactor processes, but the effects were reported to be partially reversible by return to normal operating conditions and complete eventual recovery was observed [14]. Sulphide concentrations in excess of 200 mg l^{-1} in a digester at 35 °C, with continuous feeding and mixing, produced severe inhibitory effects including the complete cessation of gas production [15] although loss of gas production was not reported to be a linear function of sulphide concentration. The reduction of the levels of soluble sulphide in the system was primarily the result of loss of H_2S with the off-gases. An equation was formulated, relating the functions of soluble sulphide reduction, digester pH, H_2S solubility and gas production, although its application was noted to be limited to those systems in which the availability of soluble sulphides was linked to the solubility of metals. Where heavy metals are present, the soluble sulphide concentration in an anaerobic digester is dependent to a marked extent on the capacity of these metals as precipitants. All the heavy metals, with the exception of chromium, form insoluble sulphide salts and can thus be removed from solution by any sulphide present in the system. Free sulphide can also be eliminated as H_2S, by vigorous gas production.

5.3 Ammonia-Nitrogen Inhibition

Although ammonia is an important buffer in anaerobic digestion processes, high ammonia concentrations can be a major cause of operational failure. Ambiguities exist in regard to the degree of toxicity ascribed to ammonia; using pilot plant experimental data and documentary evidence, the following were observed [13]:

1) operational stability was not detrimentally affected by ammonia and nitrogen concentrations in excess of threshold levels;
2) system accommodation of very high concentrations of free ammonia was not accounted for by consideration of the mechanism of cation antagonism to the ammonium ion; and,
3) balanced digester conditions were enhanced to a marked degree by initial acclimation procedures.

Hydrogen ion concentration (pH) in the liquid phase of the anaerobic digester is a function of the relative concentrations of free ammonia and the ammonium ion:

$$NH_3 \leftrightharpoons NH_4^+ + OH^-. \tag{26}$$

As pH decreases (i.e. as H^+ concentration increases), equilibrium shifts to the right. The dissociation constant of the system is 1.85×10^{-5} at 35 °C [13].

Process instability at elevated ammonia loading rates is generally considered to be the result of inadequate acclimation of the methanogenic population of the digester to increased concentrations of ammonia. In a digester system that has not previously been acclimated to high ammonia loadings, or in a reactor operating near the limits of its design capacity for anaerobic degradation, shock loadings of high ammonia concentration generally cause rapid production of VFAs such that the buffering capacity of the system may not be able to compensate for the decrease in pH. Further depression of alkalinity and reduction of pH may result in digester failure. In a digester under steady-state conditions and acclimated to high ammonia loading, relatively high concentrations of VFAs will be counter-acted by the ammonium-bicarbonate buffering capacity of the system. Acclimation of the methanogenic population to elevated ammonia concentrations is paramount, therefore, as these organisms are remarkably sensitive to pH fluctuations. However, once acclimated, an anaerobic reactor is capable of remaining in equilibrium under transient shock loading conditions.

In anaerobic digester operations, ammonia is frequently assayed as ammonia-nitrogen and in general there is no distinction attempted between ammonia and the ammonium ion. Inhibition indicated by a decrease in digester gas production and an increase in volatile acid formation appears to begin at around $1\,800$ mg l^{-1} of ammonia-nitrogen. The complete inhibition of digestion at higher concentrations does not seem to have been proved conclusively as the investigations documented may not have allowed time for complete system breakdown to occur.

The response of methanogens to concentrations of $24\,000$ mg l^{-1} of ammonium at exposure times of 1 h, 1 day and 4 days was found to be highly reversible, with rapid recovery upon removal of the ammonium from the system [14]. It is generally accepted that high levels of free ammonia under conditions of anaerobiosis are more inhibitory to the digestion process than the ammonium ion itself. The degradation of swine manure in one investigation indicated that increasing inhibition was manifested as the ammonia-nitrogen concentration exceeded $2\,000$ mg l^{-1}, but loss of gas production did not occur even at concentrations as high as $7\,000$ mg l^{-1} [13].

The inhibitory effects of ammonia, as far as is known, influence only the phase of methanogenesis in the anaerobic reactor, although other sensitive reactions, such as those performed by the OHPA bacteria, may also be either directly or indirectly affected. Ammonia-nitrogen concentrations in the range of $1\,500$ to $3\,000$ mg l^{-1} were observed to cause inhibition of methanogenic microorganisms at alkaline pH values; at concentrations in excess of $3\,000$ mg l^{-1}, the ammonium ion was claimed to be toxic irrespective of pH, and free ammonia caused toxicity at concentrations greater than 150 mg l^{-1} [11]. The methanogen *Methanobacterium formicicum* was reported to be partially inhibited in the presence of a total ammonia concentration of $3\,000$ mg l^{-1} and a pH of 7.1, with some loss of growth and CH_4-forming capacity, whereas $4\,000$ mg l^{-1} completely inhibited the microorganism. The efficient functioning of non-methanogenic anaerobic bacteria has been noted at concentrations of ammonia in excess of $6\,000$ mg l^{-1} and a pH value of 8.0 [16].

Thus, although ammonia is inhibitory to the methanogenic phase of digestion, the effects are reversible and may be avoided to a certain extent by digester acclimation.

5.4 Heavy Metals

5.4.1 The Effect of Heavy Metal Speciation in Anaerobic Digestion

The most common causative agents of inhibition and failure of sewage sludge digesters are reported to be contaminating heavy metals [1, 2]. Wastestreams from certain industrial processes containing relatively high concentrations of metals in different forms thus tend to have a degenerative effect on digester stability and performance. Heavy metals in the soluble state are in general regarded to be of more significance to digester toxicity than are insoluble forms [17, 18]. According to Mosey [19], failure of the digester system may result when the free concentration of heavy metals rises beyond a particular threshold concentration. This threshold concentration appears to be directly related to the sulphide ion (S^{2-}) concentration of digesting sludge.

Heavy metal ions can be removed, or their concentrations reduced, by several mechanisms. These include the chelation of soluble metal forms by ligands, which may be organic or inorganic. The free ions exist in a state of equilibrium in the complex and to prevent the inhibitory effects of these, the metal-chelate must possess an inherent degree of stability. The precipitation of heavy metals as salts which are not inhibitory to bacterial metabolism exists as a second mechanism of removal which can be highly efficient in some systems. Separation of solids from the liquid phase of the digester may be necessary if the metals concentration and precipitation is excessive, as a large quantity of solid matter can cause clogging of the influent and effluent pipes whilst sedimentation can cause the effective reactor volume to be reduced. The sulphide salts of the majority of the heavy metals, with the notable exception of chromium, are insoluble and can thus be removed. The carbonate salts of some metals at specified pH values (e.g. > pH 7.7 for zinc; > pH 6.4 for iron) are also poorly soluble and capable of precipitation upon formation. The latter mechanism is of particular importance where iron is concerned [19], as ferrous sulphide can be transformed to the carbonate, thus releasing the sulphide to participate in other reactions, without the simultaneous release of the metal ions into solution. The reduction of sulphates present in the wastestream provides sulphide for precipitation reactions, as does the degradation of proteins and peptides which contain the sulphydryl groups of the amino acids methionine and cysteine.

Anaerobic digestion also reduces the valency states of some metals; both copper and iron may be reduced from the tri- to the divalent state. This reduces the quantity of the precipitating agent, such as sulphide, necessary for the removal of the metal ion from solution.

As heavy metals can be removed from anaerobic systems by adsorption, those digesters such as the CSTR configuration which are tolerant of wastes containing high levels of suspended solids are effective in metal removal, provided sufficient adsorption sites are present on the solids comprising the insoluble fraction of the influent.

The effect of heavy metals on the digestion process has been shown to depend upon the form or species in which the metals are introduced into the system. The addition of nickel as nickel sulphate at 277 mg l^{-1} metal was reported to produce no alteration in digestion [20], whereas 30 mg l^{-1} nickel as the nitrate salt was observed to cause an 80% reduction of gas production [17]. The addition of nickel chloride at 500 mg l^{-1} nickel to an anaerobic filter reactor produced only a negligible effect on the evolution of biogas, but 1 000 mg l^{-1} was observed to cause severe retardation of gas production [21].

Parkin and Speece [14] reported that recovery of an anaerobic filter from the effects of nickel toxicity was largely dependent upon the metal concentration in relation to the exposure time of the system. At exposure times of less than 1 day, nickel concentrations of up to 800 mg l^{-1} were reported to result in little decrease in gas production, but gas production was halted when 2 400 mg l^{-1} nickel was added to the system for a period of 1 h. The reversibility and recovery from the effects of nickel toxicity evinced by the digester organisms, coupled with the ability of the anaerobic filter to pass the toxicant with minimal disruption to system stability were therefore considered to be functions of the metal concentration and the duration of the period of exposure. These investigations highlighted the increased tolerance of fixed-film systems in relation to those digesters dependent upon suspended growth, such as the contact and CSTR processes. The inherent capacities of increased biomass concentrations obtainable and short HRTs which can be maintained in biofilm reactors ensure less exposure time to the inhibitor. The high rates of recycle employed in the expanded and fluidised reactor bed designs enhance this effect, diluting the influent wastestream and thus protecting the digester microflora to a significant degree from the effects of inhibition. The mechanism of nickel toxicity does not appear to have been fully established. Anaerobic bacteria possess a high-affinity system for the transport of this metal [22]; nickel can hence be assimilated from the medium at extremely low concentrations. The presence of the transition metal in significant quantities is probable in reactor vessels constructed of stainless steel, as nickel is a constituent of this alloy.

Copper has also been demonstrated to inhibit the digestion process when present as its sulphate salt at concentrations of 200–230 mg l^{-1} [20]; however, when present as copper hydroxide, with 520 mg l^{-1} as copper, the effects were reported to be negligible [23]. At concentrations of copper below 200 mg l^{-1}, the lag period of growth of the methanogenic bacterium M. formicicum was observed to increase; at a total copper concentration of 300 mg l^{-1}, complete inhibition occurred, although a concentration in excess of 200 mg l^{-1} metal was required before severe inhibition of the methanogen was manifested [11].

Chromium is distinctive in that many of its salts, notably the sulphate, are soluble. When chromium was introduced into a digestion system at 100 mg l^{-1} in the trivalent state [Cr(III)], a reduction in digester gas production of 80% was recorded; however, the metal as Cr(IV), at an increased concentration of 420 mg l^{-1}, was observed to cause only slight reductions in the evolution of biogas.

The reasons for the variable effects of heavy metals and metal salts upon anaerobic digester operations are not clear. One explanation, suggested by Hayes and Theis [17], is that the concentrations of precipitating agents such as sulphide and carbonate vary from system to system, thus altering the quantities of heavy metals that may be removed from solution. Similarly, variations in the concentra-

76

tions of chelating agents in various wastestreams probably occur. The relative solubilities of various salts of the same metal and the nature and strength of adsorption to organic materials may also affect the degree of toxicity of the metal to the system.

5.4.2 The Effect of Heavy Metals on the Bacterial Flora of Anaerobic Digesters

Fewer species of bacteria have been identified in anaerobic digester systems than exist in comparable aerobic processes. The susceptibility of the former to toxic inhibition, therefore, is generally greater than that evident in aerobic activated sludge or trickling filter processes [25]. The most frequent manifestations of heavy metals inhibition in anaerobic digestion are a reduction in the evolution of biogas and the concomitant increase of VFAs concentration [26]. The methanogenic bacteria have a fastidious metabolism and were for this reason initially believed to be the major cause of digester sensitivity to toxic substances [27]. The stability of anaerobic processes is also, however, significantly dependent upon the presence and activity of another extremely sensitive group of anaerobes, the OHPA bacteria. These microorganisms grow in close proximity to the methanogens, they are intolerant of increased partial pressures of H_2 [28] and, for complex digester substrates, oxidize the fatty acids from which over half of the reactor CH_4 is produced [29].

Many wastes, such as primary and secondary sludges, mining effluent and ore processing streams, contain relatively high concentrations of heavy metals. The specific toxic effects of these on anaerobic microorganisms has not been clearly elucidated, but interference in transmembrane and other energy potentials, electron transport chains, and substrate translocation systems by means of the disruption of metabolic enzyme activities or changes in ionic gradients may occur. Several enzymes, including alcohol dehydrogenase and coenzyme A, are susceptible to inactivation by heavy metal ions, which react with the enzymic sulphydryl ($-SH$) groups. The reversibility of heavy metal toxicity often associated with anaerobic digestion [14] is indicative of noncompetitive enzyme inhibition, whereby the inhibitor combines rapidly and reversibly with the enzyme or enzyme-substrate complex at a functional group located outside the active site that is necessary for the maintenance of the catalytically active three dimensional conformation of the enzyme molecule. Catalytic activity in some enzymes is dependent to a great extent on the sulphydryl groups of certain cysteine residues of the molecule and their alteration by heavy metal ions such as mercury (Hg^{2+}) and silver (Ag^+), which form mercaptides (see Fig. 14), often causes the inactivation of the enzyme. The activity of mercury ions is easily reversed by compounds containing the sulphydryl group. Bacteria killed by silver have been found to contain 10^5 to 10^7 Ag^+ ions per cell [30].

The effect of heavy metal ions on anaerobic metabolism is an important aspect of the investigation into the overall inhibition of the digester system by heavy metals and their salts. Relevant comparisons between the degrees of toxicity of heavy metal ions and metal compounds or complexes are difficult to establish, as toxic limits to various compounds of the same metal, as noted above, are fre-

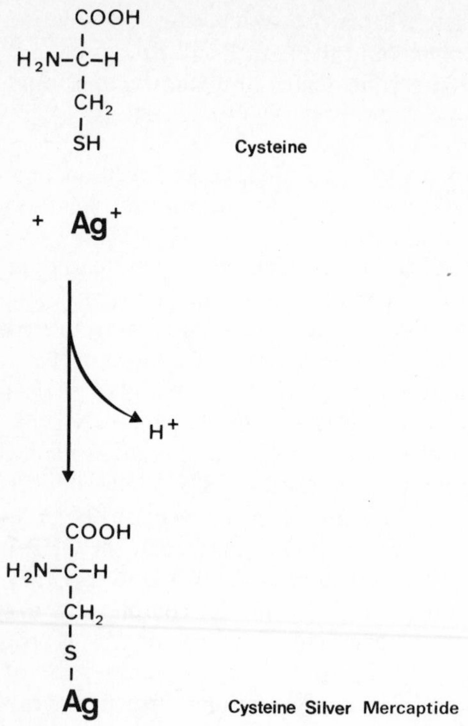

```
       COOH
         |
 H₂N–C–H
         |
        CH₂
         |
        SH            Cysteine


   +  Ag⁺



               H⁺



       COOH
         |
 H₂N–C–H
         |
        CH₂
         |
         S
         |
        Ag        Cysteine Silver Mercaptide
```

Fig. 14. Reaction of cysteine with Ag^+.
(From Lehninger A: Biochemistry, Worth Publishers Inc. 1975[2])

quently found to differ widely. The tolerance of the bacterial cell to different concentrations of heavy metal may be affected by factors other than the quantities of precipitating or chelating agents present, or the chemical properties of the metallic salt: variable reactor operational parameters such as feed concentrations, temperature and detention times within the system can affect the microflora. Reactor design may also be an influential factor: contact time between the bacterial cell within the protective extracellular biofilm and the influent inhibitor is reduced to a minimum in the high biomass fixed-film processes.

Repeated dosing with heavy metals may induce subsequent acclimation and therefore an increase of tolerance among the microbial species present. As has been reported elsewhere (Chap. 2), various wastestreams give rise to different populations of microorganisms within the reactor [11]. These differences can produce variable levels of tolerance to heavy metal dosing [17]. Another factor leading to major discrepancies in documented levels of heavy metal toxicity in the anaerobic digestion process, according to Kugelman and Chin [31], is the lack of investigation applied to the effects of synergism, antagonism and stimulation in microbial systems. The generation of CH_4, for example, was demonstrated to increase in the presence of iron concentrations of 0.2–2.0 mM. The effect was enhanced when sulphate ions were also present [32, 33]. Certain other metals, such as nickel, tungsten and selenium were also found to stimulate the growth of many anaerobic species when present in trace amounts; these observations resulted in the discovery of the metal-requiring enzymes of anaerobes [34–36].

Resistance to the effects of heavy metals is conferred on some microorganisms by the possession of plasmids or Resistance (R)-factors. The advantages of this to survival in nature is unclear [37], and the effect of the presence or transmissibility of such R-factors or plasmids on anaerobic digestion processes does not appear to have been assessed.

Mosey [19] developed a formula whereby the total metals concentration, presented on an equivalent weight basis relative to the concentration of solids in crude sewage, could be used to predict the inhibition and failure of a digester system. The total metal load constant, K, is expressed as $meqkg^{-1}$:

$$K = \left[\frac{\dfrac{Fe}{27.9} + \dfrac{Zn}{32.7} + \dfrac{Ni}{29.4} + \dfrac{Pb}{103.6} + \dfrac{Cd}{56.2} + 0.67\dfrac{Cu}{31.8}}{\text{solids concentration}} \right]. \tag{27}$$

The factor 0.67 compensates for the partial reduction of copper to the cuprous form. Mosey [19] suggested that the probability of digester failure existed at a K value in excess of 400 $meqkg^{-1}$, and that failure was almost certain when K exceeded 800 $meqkg^{-1}$.

5.5 The Effect of Cyanide

Cyano compounds can comprise a significant proportion of a number of industrial wastestreams (see Chap. 11), and their effects on anaerobic biodegradation systems are therefore of consequence. The extent of toxicant reversibility in many systems has been demonstrated to be a function of both exposure time and toxicant concentration, and cyanide is no exception [14]. Metabolic inhibition by cyano compounds is partially reversible: at levels of less than 10 mg l^{-1}, and a period of exposure of 1 h, the decrease in biogas production of an anaerobic filter system was negligible [14]. When dosage was increased to 100 mg l^{-1}, with 1 h exposure, gas production was considerably reduced but not prevented. The same dose maintained for 24 h did, however, completely inhibit gas production on the ensuing 5 days, after which time the system recovered.

The phosphorylation reaction associated with nitrate reduction caused by electron flow from NADH, i.e. oxidative phosphorylation, is intolerant of concentrations of cyanide as low as 0.1 mM. The cyanide may bind reversibly to some part of the system; it is known to inhibit electron flow. Despite the toxic effects of cyanide upon microbial systems, rapid acclimation has been recorded [16], but the consequences for anaerobic digester systems must remain ill-defined until the mechanisms of inhibition in anaerobes are fully established. Thauer and Morris [38] cite the findings that cyanide prevents CH_4 formation from acetate but not from carbon dioxide or methanol in cultures of *Methanosarcina barkeri*. This methanogen possesses carbon monoxide dehydrogenase, an enzyme which is inhibited by cyanide.

Reversibility of the effects of toxicity of cyanide and other inhibitors of bacterial metabolism must necessarily depend upon the enzyme or other system being affected. In metabolic processes which are essential to the maintenance of cell integrity, disruption by an inhibitor may have immediate and irrepairable consequences, whereas in cell systems which are not prerequisite for short term viability, or which can be by-passed by less efficient pathways, the effects of toxic materials will be of less consequence, and thus more amenable to reversibility.

5.6 Anthropogenic and Recalcitrant Compounds in Anaerobic Digestion

There is a multiplicity of natural and synthetic molecules in the environment that are known to persist unchanged for years. Many of these xenobiotic, aromatic organic chemicals are components of industrial wastestreams and many have resisted degradation under in vitro conditions. These compounds include the organochlorine insecticides such as aldrin, lindane and DDT, polymers such as polyethylene, polyvinyl alcohol and cellulose acetate, alkylbenzene sulphonate surfactants and polychlorinated biphenyls. The most persistent of the organic pesticides are generally considered to be the chlorinated hydrocarbons and certain of their associated epoxides [39, 40]. The persistence of these molecules is enhanced by their extremely hydrophobic nature which allows them to concentrate on particulate matter. The effluents of the pharmaceutical, dye-manufacturing and dyeing, tannery and pulp-mill industries are among those which contain high levels of aromatic and other organic compounds. Those recalcitrant molecules occurring naturally include many components of humic matter, such as the lignins, tannins and melanins, which are polyaromatic entities of weights of up to 2×10^6 daltons. These may have half-lives which can be measured in terms of centuries [41, 42].

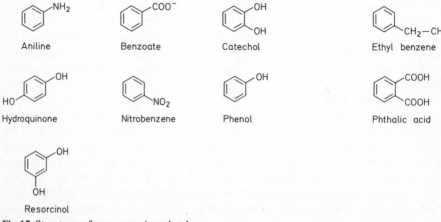

Fig. 15. Structures of some organic molecules

The difficulties apparent in the degradation and removal of many recalcitrant materials in bacterial systems may have several bases. The excessive size of these molecules, many of which contain condensed aromatic ring structures (see Fig. 15) require the combined activities of a variety of microbial species for breakdown into substituents suitable for microbial assimilation. Furthermore, the form of substituent groupings and the existence of unusual linkages and bondings are properties which tend to make large organic molecules less than amenable as microbial substrates [21], as novel enzyme systems may be necessary to attack these successfully. Organic molecules such solvents and phenolic compounds are, in addition, inhibitory in many cases to bacterial systems and enzymes [21, 43–45].

5.6.1 Response of Digester Systems to Complex Organics

The anaerobic biodegradation of large organo-compounds is generally the result of either catabolic activity by a particular microbial species in the reactor population, or the concerted catabolism of consortia of various bacterial groups. The outcome of the degradative process can be total conversion of the organic matter to cell mass and final reduction products such as CH_4 or CO_2, or only partial metabolism of the material to smaller and possibly less toxic substances. The extensive manufacture and utilisation, both industrial and domestic, of synthetic chemicals as pesticides, herbicides, refrigerants, lubricants and solvents has resulted in increased toxic loadings of manufacturing and other wastestreams, which must be dealt with efficiently in treatment systems.

Microbial growth in activated sludge systems for example, is inhibited to a marked extent by industrial wastes which contain significant proportions of mono- and dichloroaromatics [46]. In addition to the inhibitory effects, partial degradation of the aromatic constituents of these materials have resulted in increased dissolved organic carbon (DOC) concentrations and highly-coloured effluents in settlement tanks [47]; the resistant dark-coloured polymers are the result of the oxidative polymerization of polyhydric phenols in the system.

Other investigations have shown that almost no degradation of the organic colourants anthraquinone and some azo dyes occurred in activated sludge, whereas under anaerobic conditions the azo dyes were reduced to colourless constituents believed to be the corresponding amines. Degradation was reported to have been mediated by various microorganisms possessing relatively non-specific enzymes [48].

The degradation of some chlorinated organic pesticides, including aldrin, lindane and dieldrin (see Fig. 16) was observed to occur more rapidly under anaerobic conditions than under corresponding aerobic conditions [39] and degradation products that could be extracted from the system were more commonly obtained under anaerobiosis. With the exception of lindane, an increase in temperature from 20–35 °C in the operating conditions produced no significant increase in the rates of degradation of the pesticides.

Solvents in extensive use in industry and commerce are frequently chlorinated C1 and C2 hydrocarbons; these include trichloroethylene, perchloroethylene,

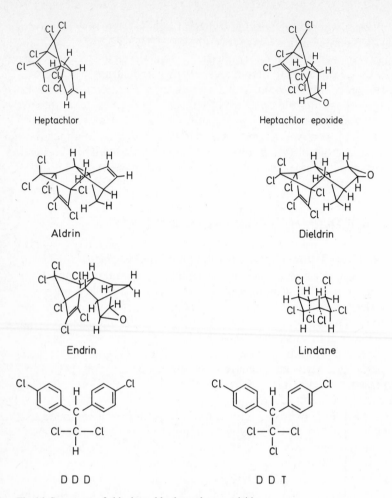

Fig. 16. Structures of chlorinated hydrocarbon pesticides

1,1,1-trichloroethene and dichloromethane. The chemical intermediates vinyl chloride and 1,2-dichloromethane also belong to this category [46]. The anaerobic biodegradation of these chloroalkanes has not been extensively studied; the complete conversion to cellular carbon and CO_2 has only been observed, as far as is known, for dichloromethane (CH_2Cl_2) [49], a widely used water-soluble component of cleaning, refrigeration and fumigation fluids. Treated sewage effluents have been found to contain up to 2 g l^{-1} CH_2Cl_2 [50].

A review on the effects of herbicides and pesticides [51] reported that soil microorganisms were not detrimentally affected in the long term by the application of chemicals at the concentrations commonly employed in agriculture; the extent to which this observation can be applied in anaerobic digestion systems, where concentrations would be naturally much greater when treating industrial wastes, is unclear. Biocidal or biostatic effects are liable to result from heavy loadings of

82

Table 8. Anaerobic degradation of recalcitrant and anthropogenic organic compounds

Compound	Ref.
Aldrin (unsaturated cyclodiene)	Hill and McCarty [39]
Aniline	Lin Chou et al. [21]
Anthraquinone	Meyer [48]
Azo dyes	Meyer [48]
Benzaldehyde	Williams and Evans [71]
Benzoate	Tarvin and Buswell [57] Proctor and Scher [72] Dutton and Evans [73, 74] Bakker [63] Healy and Young [75] Lin Chou et al. [21] Mountfort and Bryant [76] Sleat and Robinson [77]
Benzyl alcohol	Williams and Evans [71]
Bromochloromethane	Lang et al. [78]
Caffeic acid	Williams and Evans [71]
Catechol	Healy and Young [68, 75] Balba et al. [79] Lin Chou et al. [21]
4-Chlorobiphenyl	Sylvestre and Fauteaux [80]
Chloroethane	Lang et al. [78]
Cinnamate	Williams and Evans [71] Evans [61] Healy and Young [75]
o,m,p-Cresol	Bakker [63, 81] Harper et al. [45]
DDD	Hill and McCarty [39]
DDT (1,1'-bis(p-chlorophenyl)-2,2,2-trichloroethane)	Guenzi and Beard [82] Hill and McCarty [39] Ware and Roan [83] Pfaender and Alexander [84] Esaac and Matsumara [53]
Dibromochloromethane	Lang et al. [78]
Dichloroethane	Lang et al. [78]
Dichloromethane	Rittmann and McCarty [49]
Dieldrin (epoxidised cyclodiene)	Hill and McCarty [39] Batterton et al. [85] Esaac and Matsumara [53]
Dihydroxybenzene	Schink and Pfennig [82]
Dimethylnitrosamine	Prins [86] Kobayashi and Tchan [87]
Endrin (epoxidised cyclodiene)	Hill and McCarty [39]
Ethyl benzene	Lin Chou et al. [21]

Table 8 (continued)

Compound	Ref.
Ferulic acid	Healy and Young [75]
Heptachlor (unsaturated cyclodiene)	Hill and McCarty [39]
Heptachlor epoxide (epoxidised cyclodiene)	Hill and McCarty [39]
Heptachlorobornane	Esaac and Matsumara [53]
Hydroquinone	Lin Chou et al. [21]
m-Hydroxybenzoate	Dutton and Evans [74] Taylor et al. [88] Williams and Evans [71] Bakker [81]
o-Hydroxybenzoate	Taylor et al. [88] Balba et al. [79]
p-Hydroxybenzoate	Dutton and Evans [74] Taylor et al. [88] Williams and Evans [71] Bakker [81] Balba et al. [79] Healy and Young [75]
p-Hydroxycinnamate	Williams and Evans [71]
Kepone	Allen et al. [89]
Lindane (benzene hexachloride isomers)	Hill and McCarty [39] Ware and Roan [83] Esaac and Matsumara [53]
o,m,p-Methoxybenzoate	Balba et al. [79]
Methoxychlor	Paris and Lewis [90]
Methyl chloride	Lang et al. [78]
Methylene chloride	Lang et al. [78]
Nitrobenzene	Lin Chou et al. [21]
p-Nitrophenol	Esaac and Matsumara [53]
Parathion	Esaac and Matsumara [53]
Pentachlorophenol	Chu and Kirch [91] Esaac and Matsumara [53]
Phenol	Bakker [63, 81] Healy and Young [68, 75] Balba et al. [79] Lin Chou et al. [21] Cross et al. [16] Harper et al. [45]
Phenylacetate	Tarvin and Buswell [57] Williams and Evans [71] Balba and Evans [92]
Phenylalanine	Evans [61]
Phenylpropionate	Tarvin and Buswell [57]
Phloroglucinol	Evans [61]

84

Table 8 (continued)

Compound	Ref.
Phorate sulphoxide	Esaac and Matsumara [53]
Phthalate	Lin Chou et al. [21]
	Aftring et al. [93]
	Aftring and Taylor [94]
Phthalate esters	Aftring et al. [93]
Protocatechuate	Taylor et al. [88]
	Williams and Evans [71]
	Bakker [63, 81]
	Balba et al. [79]
	Healy and Young [75]
Quinate	Evans [61]
Resorcinol	Lin Chou et al. [21]
Shikimate	Evans [61]
Sodium benzoate	Sleat and Robinson [77]
Syringaldehyde	Healy and Young [75]
Syringic acid	Healy and Young [75]
	Kaiser and Hanselmann [95]
Tetrachloroethylene	Bouwer et al. [96]
	Lang et al. [78]
Toluene	Aftring et al. [93]
Toxaphene	Paris and Lewis [90]
Trichloroethane	Lang et al. [98]
Trichloroethylene	Bouwer et al. [96]
	Lang et al. [78]
Trichloromethane	Bouwer et al. [96]
	Lang et al. [78]
Trihydroxybenzene	Schink and Pfennig [64]
Vanillic acid	Healy and Young [75]
	Zeikus et al. [70]
Vanillin	Healy and Young [75]
Vinylidiene chloride	Lang et al. [78]

pesticides and other large organic molecules, unless biodegradative pathways exist among the microbial enzyme systems of the reactor flora; the size of these organic molecules and their charge patterns at least may effectively prevent microbial enzyme activity by steric interference. Several compounds have proved to be toxic to acetate-grown cultures of anaerobic bacteria [21]. These included 3-chloro-1,2-propanediol, 2-chloro-propionic acid, 1-chloro-propane, 1-chloro-propene, acrolein and formaldehyde, at concentrations of 500 mg l^{-1} (see Table 8). The addition of ethyl benzene and aniline proved non-toxic to these cultures.

Formaldehyde concentrations of 75 mg l^{-1} were observed to result in 50% inhibition of enriched anaerobic cultures, but an anaerobic filter was tolerant of 500 mg l^{-1} formaldehyde for up to 4 days, with little inhibition of digestion evinced [21], although 7 days continuous feed at this dose caused the cessation of gas production. Methanogenesis in the same system was retarded by 500 mg l^{-1} of 2,4-dinitrophenol. Another extremely persistent compound, pentachlorophenol, a widely used wood preservative and fungicide, has been reported to be extremely toxic to anaerobic digestion processes; a toxic concentration in sewage of as low as 0.4 mg l^{-1} was recorded [43]. The pharmaceutical products dichlorophen and hexachlorophane are also both inhibitory to sewage treatment processes at low levels, as is 2-mercaptobenzothiazole [44]. Anthony and Breimhurst [52] listed a number of pollutants toxic to anaerobic digestion, including acrylonitrile, benzidine and carbon tetrachloride.

The mechanisms of inhibition of microbial metabolism in the above systems are not clear; the manifestations of toxicity are generally loss of gas production in the digester and accumulation of intermediary metabolites. As the microbial division cycle may be suspended for a period when a toxic substance is introduced into the system, the maintenance of an extended SRT together with a short HRT to ensure process stability is of fundamental importance [14, 21].

5.6.2 Response of Anaerobic Bacteria to Hazardous Organic Molecules

In contrast to aerobic processes, the anaerobic biological degradation of hazardous waste does not require the continuous aeration of the bacterial system, nor does the process produce the large quantities of sludge associated with aerobic systems. Anaerobic bacterial processes are hence energy-efficient, and those reactions such as the reduction of nitrates and sulphoxides, and reductive dehalogenations that are frequently necessary in the degradation of complex recalcitrant substrates may be enhanced [53]. According to Kobayashi and Rittmann [40], two limitations arise in the study of biodegradative phenomena. The first entails the use of high concentrations of the recalcitrant organic substrate in laboratory investigations of biodegradability; in many situations, only $\mu g\, l^{-1}$ of compound in the influent may be required to be reduced in the effluent. These trace amounts may not be sufficient to induce the necessary catabolic enzyme systems [54–56]. The second limitation involves the inhibition of biodegradation by the antagonistic interactions of microorganisms [40]. Additionally, in digester systems, the preferential utilisation of other more freely available substrates may preclude the catabolism of complex organic molecules, if the latter do not exert inhibitory effects on the microflora of the system. However, axenic and nonaxenic cultures of bacteria have been utilised in the study of anaerobic degradation and its various metabolic pathways, and several of the phenomena involved have been elucidated.

The biological degradation of benzoid structures under strictly anaerobic conditions was first observed to produce CH_4 and CO_2 via the activity of an accli-

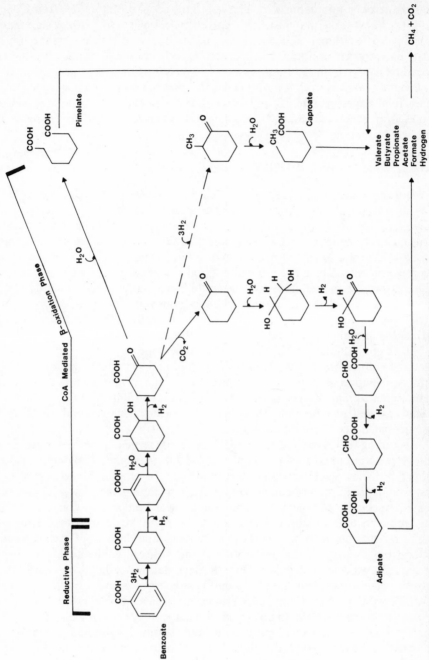

Fig. 17. Pathways in the anaerobic degradation of benzoic acid. (After Evans [61])

mated microbial population [57]. This led to the conclusion that this transformation was likely to be ubiquitous in natural environments and in the conversion of many waste effluents and pesticides. The methanogenic bioconversion of benzoate has been shown to occur by reduction of the aromatic ring to chlorohexanecarboxylate (see Fig. 17); this is succeeded by a β-oxidation sequence mediated by coenzyme A, an enzymic acyl carrier. The alicyclic ring is thus cleaved to yield in the final stages, acetate, H_2 and CO_2 – the substrates of the methanogenic bacteria – from which CH_4 and CO_2 are the resultant end products, according to the following equation (58):

$$4C_6H_5CO_2H + 18H_2O \rightarrow 15CH_4 + 13CO_2. \tag{28}$$

A stabilised consortium of anaerobic microorganisms perform the conversion [58, 59] and post-ring-rupture intermediates have been tentatively identified as heptanoic, valeric, butyric, propanoic and acetic acids [60]. The reduction and cleavage of the benzene nucleus is mediated by Gram-negative facultative organisms [61]. The microbial consortium involved in one methanogenic conversion of benzoic acid was found, upon subculture, to include a facultative Gram-negative bacterium and two methanogenic species [62]. The latter could not utilise benzoate alone. A facultative Gram-negative organism present in another consortium investigated could degrade benzoate anaerobically in the presence of nitrate [61]. Other alicyclic compounds were also utilised.

The degradation of phenol under anaerobic conditions by a heterogeneous bacterial culture originating from soil, manure and sewage sludge was demonstrated [63]; upon serial transfer in liquid culture, the composition of the consortium remained almost unchanged, consisting of a spirillum and Gram-negative organisms. Benzoate, monohydroxybenzoates, protocatechuate and cresols were also degraded (see Table 8).

According to Thauer and Morris [38], benzoate fermentation to acetate, CO_2 and H_2 under standard state conditions (pH 7.0) is an endergonic reaction, although benzoic acid conversion to CH_4 and CO_2 is exergonic. Hence, the former degradation requires syntrophic associations including methanogenic or sulphate-reducing bacteria which can maintain low H_2 and acetate concentrations. Anaerobic bacteria capable of the fermentation of trihydroxybenzenes were isolated from marine and limnic mud samples: the greater the level of hydroxylation of the aromatic ring of benzene derivatives, the more exergonic the acetate-forming reaction was found to be [64]. Only the degradations of dihydroxy- and trihydroxybenzenes were concluded to be sufficiently exergonic to support monocultural bacterial growth. Some sulphate-reducing bacteria can however, oxidise benzoic acid, via sulphate, to obtain carbon and energy for growth [38].

A number of organic compounds, including some chlorinated hydrocarbon pesticides, appear to be more rapidly degraded by a complex population of bacteria than by a homogeneous culture under anaerobic conditions [39, 65]. The reaction of a bacterial digester population to the presence of a recalcitrant material must depend to some extent upon environmental parameters such as temperature and pH, both of which affect enzymic activity; the latter may also influence the state of the toxicant, rendering it liable to precipitation or dissociation.

However, the structure of the refractory molecule is the major impediment to bacterial catabolism. This can be overcome by various means; the cometabolism of a molecule by one species of digester microbe may release a substrate suitable for another [66]; gene mutation could furnish an organism with a novel degradative capacity, which may be further transmitted in the population by transposon or plasmidial means. This latter may be one of the mechanisms of the acclimation phenomena frequently encountered in anaerobic digester populations exposed to a novel compound.

Adaptation to the toxicant or recalcitrant is an important consideration in microbial systems: the reactor bacteria possessing enzymes that are capable of the catabolism, partial or complete, of the alien molecule, may be slow growing or extremely sensitive and appear later in the digestion process. Anaerobic cultures have been reported to recover from shock doses of toxicants after several months have elapsed [67]. Under conditions of the strictest anaerobiosis, the biodegradability of eleven aromatic derivatives of the extremely complex heteropolymer lignin was demonstrated [68]. Acclimation procedures carried out using a heterogeneous microbial population and a particular aromatic substrate resulted in the bacteria simultaneously acclimating to other aromatic substrates possessing similar substituent group arrangements on the aromatic ring.

The above observation is an indication of the remarkable versatility of microbial enzyme systems, which enables the organisms possessing them to deal with the naturally occurring aromatic derivatives of lignins, terpenes and tannins. The degradation of xenobiotic materials requires that these microbial enzyme systems recognise and accept substrate molecules with structures similar to, but not completely identical with, metabolisable natural substances. Furthermore, the synthetic or novel molecule must be capable of inducing or derepressing the synthesis of the necessary transforming enzyme or enzymes. The factors influencing the degradation of recalcitrant xenobiotic moieties in the reactor system include the structural characteristics of the novel substrate, such as molecular size, the presence or lack of particular side groups and substituents, steric factors and bonding forms and strengths. The types and numbers of available microorganisms within the anaerobic system, and their constitutive or inducible enzymes must likewise influence the degradative reactions and the extent to which they occur. In addition, natural compounds may exist at artificially high concentrations in many industrial wastestreams. These could interfere with catabolic enzyme systems and alter energy balances or membrane or other potentials. Their preferential uptake in a microbial system may also cause other essential compounds to become limiting.

The acclimation characteristics and degradation rates of various petrochemical compounds metabolised by microbial cultures were found to be affected by the structure and position of functional groups [21]. The list of petrochemicals catabolised by bacteria after acclimation procedures had been completed included acetaldehyde, catechol, vinyl acetate, nitrobenzene and phthalic acid. Those compounds with ester, hydroxyl and carboxyl groups acclimated more readily than those possessing carbonyl, amino or chloro groups, and in addition exhibited more rapid degradation rates; 2-propanol was found to have a greater utilisation efficiency than propanol.

The microbial degradation pathways of recalcitrant organics such as the ligno-aromatic compounds have been analysed and decomposition models outlined [58, 61, 69]. Investigations have been by no means exhaustive, and although lignin itself must be depolymerized or chemically modified prior to anoxic degradation [70], a diversity of these compounds are probably amenable to anaerobic breakdown to CH_4 and CO_2.

References

1. Swanwick JD, Shurben DG, Jackson S (1969) Water Pollut Control 68:639
2. Duarte AC, Anderson GK (1983) I Chem E Symp 77:149
3. Wukasch RF, Grady CPL, Kirsch EJ (1981) AIChE Symp Series 209 77:137
4. Good P, Moudry R, Fluvi P (1982) Biotechnol Lett 4:565
5. Owen WF, Stuckey DC, Healy JB Jnr, Young LY, McCarty PC (1979) Water Res 13:485
6. Parkin GF, Speece RE, Yang CHJ (1981) A comparison of the response of methanogens to toxicants: anaerobic filter vs. suspended growth systems. In: Proc seminar/workshop on anaerobic filters: an energy plus for wastewater treatment. Argonne Nat Lab, Argonne Illinois, p 37
7. D'Adamo PD, Rozich AF, Gaudy AF Jnr (1984) Biotechnol Bioeng 26:397
8. Leão C, van Uden L (1984) Biotechnol Bioeng 26:403
9. Ianotti EL, Fischer JR (1984) Effects of ammonia, volatile acids, pH and sodium on growth of bacteria isolated from a swine manure digester. In: Developments in industrial microbiology: Proc 40th Gen Meeting Soc Ind Microbiol, Sarasota, Florida, Aug. 14–19, 1983. Victor Graphics, Baltimore, p 741
10. McCarty PL, Brosseau MH (1963) Effects of high concentration of individual volatile fatty acids on anaerobic treatment. In: Proc 18th Ind Waste Conf, Purdue Univ., Lafayette, Indiana 1963. Ann Arbor Science, Ann Arbor Michigan, p 283
11. Hobson PN, Shaw BJ (1976) Wat Res 10:849
12. Andrews JF (1969) J Sanit Eng Div ASCE 95, SAI 95–0000
13. Kroeker EJ, Schulte DD, Sparling AB, Lapp HM (1979) J Water Pollut Control Fed 51:718
14. Parkin GF, Speece RE (1983) Water Sci Technol 15:261
15. Lawrence AW, McCarty PL, Guerin FJA (1964) Proc 19th Ind Waste Conf, Purdue Univ, Lafayette, Indiana 1964. Ann Arbor Science, Ann Arbor Michigan, p 343
16. Cross WH, Chian ESK, Pohland FG, Harper S, Kharkar S, Cheng SS, Lu F (1983) Biotechnol Bioeng Symp 12:349
17. Hayes TD, Theis TL (1978) J Water Pollut Control Fed 50:61
18. Sterritt RM, Lester JN (1984) Sci Total Environ 34:117
19. Mosey F (1976) Water Pollut Control 75:10
20. McDermott GN, Post MA, Jackson BN, Ettinger MB (1965) J Water Pollut Control Fed 37:163
21. Lin Chou W, Speece RW, Siddiqi RH (1979) Biotechnol Bioeng Symp 8:391
22. Jarrell KF, Sprott GD (1982) J Bacteriol 151:1195
23. Masselli JW, Masselli NW, Burford G (1961) The occurrence of copper in water, sewage and sludge, and its effect on sludge digestion: New England Interstate Water Pollut Control Commission, Boston Mass, USA
24. Moore WA, McDermott GN, Post MA, Mandia JW, Ettinger MB (1961) J Water Pollut Control Fed 33:54
25. De Walle FB, Chian ESK, Brush J (1979) J Water Pollut Control Fed 51:22
26. Kirk PWW, Lester JN, Perry R (1982) Water Res 16:973
27. Stein RM, Malone CD (1980) Environ Technol Lett 1:571
28. Verstraete W, De Baere L, Rozzi A (1981) Trib Cebedeau 34:367
29. Kaspar HF, Wuhrmann K (1978) Appl Environ Microbiol 36:1
30. Davis BD, Dulbecco R (1980 [3]) Sterilization and Disinfection. In: Davis BD, Dulbecco R, Eisen HN, Ginsberg HS (eds) Microbiology. Harper and Row, Hagerstown Maryland, p 1263

31. Kugelman IJ, Chin KK (1971) Adv Chem Ser 105:55
32. Hoban DJ, van den Berg L (1979) J Appl Bacteriol 47:153
33. van den Berg L, Lamb KA, Murray WD, Armstrong DW (1980) J Appl Bacteriol 48:437
34. Ljungdahl LG (1976) TIBS 1:63
35. Stadtman TC (1980) TIBS 5:203
36. Thauer RK, Diekert G, Schönheit P (1980) TIBS 5:304
37. Broda P (1979) Plasmids. W. H. Freeman, Oxford San Francisco, p 129
38. Thauer RK, Morris JG (1984) Metabolism of chemotrophic anaerobes: old views and new perspectives: Kelly DP, Carr NG (eds) Symp 36 (II) Soc Gen Microbiol, Univ., Warwick, April 1984. Cambridge Univ. Press, Cambridge, UK, p 123
39. Hill DW, McCarty PL (1967) J Water Pollut Control Fed 39:1259
40. Kobayashi H, Rittmann BE (1982) Environ Sci Technol 16:107A
41. Jenkinson DS, Rayner JH (1977) Soil Sci 123:298
42. O'Brien BJ, Stout JD (1978) Soil Biol Biochem 10:309
43. Drew EA, Swanwick JD (1963) Inst Publ Hlth Engrs J 62:61
44. Thom NS, Agg AR (1975) Proc R Soc Lond B 189:347
45. Harper SR, Cross WH, Pohland FG, Chian ESK (1984) Biotechnol Bioeng Symp 13:401
46. Knackmuss H-J (1983) Biochem Soc Symp 48:173
47. Haller HD (1978) J Water Pollut Control Fed 50:2771
48. Mayer U (1981) FEMS Symp 12:371
49. Rittmann BE, McCarty PL (1980) Appl Environ Microbiol 39:1225
50. Reinhard M, Dolce CJ, McCarty PL, Argo DG (1979) J Environ Eng Div ASCE 105(EE4):675
51. Greaves MP, Davis HA, Marsh JAP, Wingfield GI (1976) CRC Crit Rev Microbiol 5:1
52. Anthony RM, Breimhurst LH (1981) J Water Pollut Control Fed 53:1457
53. Esaac EG, Matsumara F (1980) Pharm Ther 9:1
54. Boethling RS, Alexander M (1979) Appl Environ Microbiol 37:1211
55. Boethling RS, Alexander M (1979) Environ Sci Technol 13:989
56. Rittmann BE, McCarty PL (1980) Biotechnol Bioeng 22:2359
57. Tarvin D, Buswell AM (1934) J Am Chem Soc 56:1751
58. Balba MT, Evans WC (1977) Biochem Soc Trans 5:302
59. Ferry JG, Wolfe RS (1976) Arch Microbiol 107:33
60. Keith CL, Bridges RL, Fina LR, Iverson KL, Cloran JA (1978) Arch Microbiol 118:173
61. Evans WC (1977) Nature, London 270:17
62. Ferry JG, Wolfe RS (1977) Appl Environ Microbiol 34:371
63. Bakker G (1977) FEMS Lett 1:103
64. Schink B, Pfennig N (1982) Arch Microbiol 133:195
65. Raghu K, Macrae IC (1966) Science 29:263
66. Horvath RS (1972) Bacteriol Rev 36:146
67. Parkin GF, Speece RE (1982) J Environ Eng Div ASCE 108(EE3):515
68. Healy JB, Young LY (1978) Appl Environ Microbiol 35:216
69. Healy JB, Young LY, Reinhard M (1980) Appl Environ Microbiol 39:436
70. Zeikus JG, Wellstein AL, Kirk TK (1982) FEMS Microbiol Lett 15:193
71. Williams RJ, Evans WC (1975) Biochem J 148:1
72. Proctor MH, Scher S (1960) Biochem J 76:33
73. Dutton PL, Evans WC (1967) Biochem J 104:30
74. Dutton PL, Evans WC (1969) Biochem J 113:525
75. Healy JB, Young LY (1979) Appl Environ Microbiol 38:84
76. Mountfort DO, Bryant MP (1982) Arch Microbiol 133:249
77. Sleat R, Robinson JP (1983) J Gen Microbiol 129:141
78. Lang RR, Wood PR, Parsons RA, Demarco J, Harween HJ, Payan IL, Meyer LM, Ruiz MC, Ravelo ED (1981) Introductory study of the biodegradation of the chlorinated methane, ethane and ethene compounds, presented at AWWA annual meeting, St. Louis, Mo., June 7–11
79. Balba MT, Clarke NA, Evans WC (1979) Biochem Soc Trans 7:1115
80. Sylvestre M, Fauteaux J (1982) J Gen Appl Microbiol 28:61
81. Bakker G (1977) Degradation of aromatic compounds by microorganisms in dissimilatory nitrate reduction. Doctorate Thesis, Technical University, Delft, Netherlands
82. Guenzi WD, Beard WE (1967) Science 156:1116

83. Ware GW, Roan CC (1970) Residue Rev 33:15
84. Pfaender FK, Alexander M (1972) J Agric Food Chem 20:842
85. Batterton JC, Boush GM, Matsumara F (1971) Arch Environ Contam Toxicol 6:589
86. Prins RA (1978) Nutritional impact of intestinal drug-microbe interactions. In: Hathcock J, Coons J (eds) Nutrition and drug interrelations. Academic Press, New York, p 189
87. Kobayashi M, Chan YT (1978) Water Res 12:199
88. Taylor BF, Campbell WL, Chinoy I (1970) J Bacteriol 102:430
89. Allen GC, Young RW, Benoit RE (1981) Degradation of kepone by anaerobic sediment bacteria, presented at Am Soc Microbiol Annual Meeting
90. Paris DF, Lewis DL (1973) Residue Rev 45:95
91. Chu J, Kirsch EJ (1972) Appl Microbiol 23:1033
92. Balba MT, Evans WC (1979) Biochem Soc Trans 7:403
93. Aftring RP, Chalker BE, Taylor BF (1981) Appl Environ Microbiol 41:1177
94. Aftring PR, Taylor BF (1981) Arch Microbiol 130:101
95. Kaiser JP, Hanselmann KW (1982) Arch Microbiol 133:185
96. Bouwer EJ, Rittmann BE, McCarty PL (1981) Environ Sci Technol 15:596

6 Single-Stage Non-Attached Biomass Reactors

Several anaerobic reactor types are utilised for waste treatment by biological means; these can be broadly divided into two groups, namely the fixed-film reactors and the non-attached growth systems. The biomass of the former comprises bacteria attached as films to inert supportive media; the latter depend for their operation on the metabolic activity of microorganisms suspended as flocs or granules in the reactor vessel. The bacteria in suspended growth processes must form flocs to remain in the reactor, and the efficiency of non-attached biomass systems is to a great extent dependent upon the floc-forming and settling abilities of the sludge inoculum used to initiate the anaerobic digestion process. The microflora of anaerobic reactors are almost exclusively bacterial; although protozoa may be present, they are generally introduced with the influent and play no active part in the degradation reactions [1].

6.1 The Continuously Stirred Tank Reactor

6.1.1 Design and Operation

In common with many of the anaerobic bioreactor systems, the simple mix digester or continuously stirred tank reactor (CSTR) (Fig. 18) was developed from its aerobic counterpart. For effective treatment, this reactor design requires an extended HRT: it has no specific means of biomass retention, thus the SRT must be sufficiently high to permit biological conversion reactions to occur. The overall limitation to the rate of degradation in those digester systems operating on animal and vegetable wastes and the majority of sewage sludges is the breakdown of fibrous matter and cellulose. These types of wastestream also frequently contain high levels of suspended solids, and the single stage CSTR is generally well-

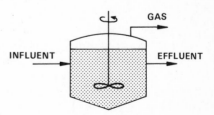

Fig. 18. The continuously-stirred tank reactor

adapted to treat them. It should be noted, however, that the minimum HRT values for fully-mixed digesters found in the literature tend to reflect the growth rates of the acetate-converting methanogenic bacteria [2].

The conventional single-stage CSTR comprises a vessel of steel, concrete or brick; several waste digesters in Europe and America have incorporated enamelled slurry-storage tanks as the basic reactor vessel [1]. Some form of plastic generally constitutes the insulation, but mineral wool has also been utilised [3]. Optimal thicknesses of digester insulation are about 10–19 cm. The HRT of the system is inherent in the rate of organic loading and varies for heated digesters from 10–60 days; for cold digesters, the HRT can range from 90–200 days [4]. The size of the reactor is estimated from the daily volume of waste to be converted and the detention time necessary for efficient treatment; the vessels are usually cylindrical in shape and vary from about 2–3000 m^3 [1] (see Fig. 18). CSTR systems generally incorporate a floating top, which also holds the gas generated in the digestion process, or a fixed top and an external gas holder. The biogas produced in the system can be stored and used at a low pressure of approximately 125 mm water gauge [5]. The Chinese digester, as described by Hobson [1] is exceptional in that it comprises an upper dome of fixed volume in which the biogas accumulates under pressure. The tank from which gas is liberated at varying pressure is situated below ground level, and is adequate for simple gas utilisation; for more sophisticated engines or gas-boilers, a constant gas pressure is necessary.

The anaerobic digester is technically a continuous microbial culture and as such requires a continuous input of medium that is balanced by a continuous outflow of digested waste and excess biomass. The influent pumps utilised to transfer slurry feedstocks are usually of too great a pumping volume to permit the continuous feeding of the small volumes necessary for long detention periods within the reactor. Input is therefore generally intermittent; timed pumps can be manipulated to give optimum performance to suit the installation. In a commercial biogas-producing plant a cyclic pumping process of 4 min of operation followed by 20 min of non-operation in a 325 m^3 capacity reactor gave optimum performance at a pumping rate of 105 l min^{-1} for digestion of slurry [5]. Generally, digestion is more efficient the shorter the intervals between non-operation of the pump. In very small digester systems, feed intervals can range from once per hour to once per day [3]. A continuous feed input by gravity flow may be possible in the smaller digester and also in larger installations if the digester is below ground. The outflow in small digesters is frequently by gravity flow and a constant head of liquid, to balance internal gas pressure, is maintained by flow over some type of weir [1]. Some large digesters use pumped output.

Mixing of the vessel contents in a CSTR process is generally achieved by paddle or screw systems or by gas diffusers (draught tubes) situated near the base of the assembly; agitation tends to be intermittent in these reactors and can be maintained by periodic recirculation of the biogas produced [5]. Mechanical agitation is frequently employed in the smaller digesters: in one investigation, mixing times of 3 min every hour and 5 min every 2 h were utilised in two small operations (average vessel volume 67 m^3) and a temperature-linked agitation system was reported [3]. Free-rising bubbles of biogas within the system may be recirculated in large reactors; some digesters depend entirely on this form of mixing, although insufficient agitation can result in the serious problem of scum formation.

The heating of the anaerobic CSTR unit can be effected by the utilisation of an external water heat-exchanger: the vessel contents are pumped through the exchanger as necessary upon a signal from a digester thermostat. This system is used on some sewage digesters, but the main disadvantage is the large pump that is required to circulate the sludge. Internal water circulation heat-exchangers are now used most frequently in commercial situations, as these are more economical and compact and require only a small pump for hot water circulation [1]. Heat-exchangers may be designed as coiled tubes or flat panel forms, but both types need efficient mixers beneath them to ensure adequate sludge circulation and to prevent clogging. Mixing and heating can be combined in double-skin draught tubes, where the hot water circulates through the double skin [1]. In a temperate climate, energy may be conserved by a transfer of heat from the digester output, but this system of heat exchange produces varied problems in construction and operation, although an experimental heat exchanger of this nature has been described [6].

A CSTR with no means of agitation is often termed a plug-flow or tubular reactor. The medium enters one end of the vessel, flows along its length and exits at the other end. An inoculum added to the influent grows as the medium flows along the vessel, utilising the available substrate; the system behaves as a batch culture and the rate of flow dictates the amount of bacterial degradation of the substrate. Each medium volume passes as a discrete mass and, theoretically, no mixing occurs in the vessel. No inoculation of fresh input material can thus be made by the reactor microflora and continuous operation of the system therefore requires continuous inoculation of bacteria. A recycle assembly can be used for influent inoculation if the feedstock has no indigenous flora. In practice, a plug-flow digester cannot be maintained, as wall friction, convection currents and gas production mix the vessel contents. The detention time of a tubular reactor must be similar to that of the conventional CSTR to allow efficient biodegradation.

6.1.2 Process Efficiency

The CSTR is the most simple of the anaerobic digesters in common use. The SRT and the HRT in this type of reactor are equal and tend to be protracted; high rate anaerobic treatments using the CSTR are therefore not possible. These systems have however been utilised successfully for the stabilisation of sewage sludge and the conversion of industrial wastewaters which contain high solids concentrations, such as crop residues [7].

The reaction of CSTR systems to unfavourable environmental conditions has been reported to be unsatisfactory. A continuously mixed reactor with an SRT of 10 days and no provision for recyle experienced digestion failure if formaldehyde was introduced at 100 mg l^{-1}; the digester ceased to produce biogas after 4 days and had not recovered after 30 days [8]. This indicates the necessity for maintenance of prolonged SRTs if process stability in the presence of adverse environmental conditions is to be ensured. The investigators also reported that the

CSTR microflora could not acclimate to or metabolise some petrochemicals (although the microbial population of anaerobic filter reactors did not have this problem), as the 20 day SRT necessary in the system tended to result in washout of the biomass.

The activities and efficiencies of ten anaerobic full-scale digesters were investigated, eight of which were of the stirred tank configuration: the loading rates employed varied from 0.7–3.2 kg VS m^{-3} d^{-1} and the wastes contained 4.7–11.3% total solids [4]. Reductions of 27–44% VS were reported and CH$_4$ production ranged from 53–70% of the total gas. Generally, superior performances were exhibited by those systems operating in the mesophilic (27–36 °C) as opposed to psychrophilic (13–21 °C) ranges; retention times were protracted, ranging from 13–80 days, the latter exhibited by a reactor operating at 13 °C. At retention times in excess of 25 days, the VS reductions obtained seemed to depend on the agitation system employed rather than the temperature. Similar VS reductions have been reported elsewhere [9, 10] although loading rates were low compared to some other similar systems [10]. The pH levels in the digesters presented no problems and did not fall below 6.9 in any system.

As the biomass in the CSTR configuration is not retained by settling or attachment, the percentage of COD removal tends to be limited [11]. In typical reactors at temperatures of 30–35 °C retention times are in the range 10–20 days, but these vary with waste composition and degree of agitation. An investigation into the effects of variable agitation rates, with decreasing retention times and increased loading rates, indicated that stability of operation was maximal at loads of up to 7.3 g VS l^{-1} d^{-1} and retention times of 3 days in the thermophilic range [12]. A retention time of 10 days at 55 °C was reported in the treatment of slaughterhouse wastes at loading rates of 9 kg VS m^{-3} d^{-1}, although at mesophilic temperatures, 20 day retention periods were necessary to treat 4.5 kg VS m^{-3} d^{-1} [12]; these observations were confirmed in another investigation, where 14 days were required to treat a maximal volumetric loading rate of 4 kg VS m^{-3} d^{-1} [14].

The proportion of anaerobic digester systems in commercial use has increased since 1982 [15] and on-site construction can take place in approximately 4 weeks [5]. A thermophilic system in use for the treatment of municipal waste was reported to process twice the volume of sludge normally associated with similar mesophilic operations, although similar COD reductions were obtained; a 7 day retention time was reported, although the steady state condition took 1 year to attain [16]. The 2–3 week retention time employed by the Hyperion treatment plant in Los Angeles, USA, enabled loading rates of 7.8 kg VSS m^{-3} d^{-1} to be applied, with COD reductions of 59% achieved [17].

CSTR systems are susceptible to malfunction upon shock loading or subsequent to the introduction of a variety of toxic substances. Malfunction manifests itself in terms of reduced gas production, reduced degradation of organic materials and simultaneous increase in acidity. Despite this, McConville and Maier [18] reported increased CH$_4$ generation and increased VSS reduction at different retention times under shock load conditions upon dosing their reactors with powdered activated carbon (PAC). In essence, these workers were converting the CSTR to a carrier-assisted process, whereby substrate and bacteria are adsorbed onto the surface of the PAC particles.

The improved digestion of animal waste by the simple expedient of gravitational settling of the contents of a conventional mixed digester prior to effluent removal has been demonstrated [19]. A 36% increase in gas yield, a 3.5-fold increase in fibre retention, a 2-fold increase in fibre breakdown and a 2.5 fold increase in the concentration of active digester bacteria in comparison with a conventional unit were recorded. The time required for this additional treatment is the major drawback, however, thus widespread adoption of the technique seems unlikely.

CSTR systems are limited by the extended HRTs necessary for efficient waste treatment, but their tolerance of high influent levels of suspended solids indicates their suitability for the stabilisation of several types of wastestream.

6.2 The Contact Process

6.2.1 Design and Operation

The contact or recycled flocs process comprises a continuously-fed, completely-mixed reactor stage followed by solids/liquid separation (see Fig. 19). A degasification step is frequently included in system design. The effluent is discharged from the settling device and the settled biomass returned to the digester vessel, where it is mixed with the incoming feed. Reinoculation of a well-acclimatised sludge can maintain optimum stabilisation of industrial wastewaters which, unlike sewage sludges for example, do not generally contain a high proportion of microflora.

The bacteria in a contact reactor occur as suspended flocs and the system is maintained in suspension by mechanical stirring, gas sparging or recycle. Inert particles in the feedstock may act as media to convert the reactor to the carrier-assisted contact process (see Chap. 7), but in general the bacteria must form flocs to remain in the system. Separation of flocs and treated wastewater occurs in a separator assembly such as a sedimentation tank (see Fig. 19) from which the suspended settled flocs are recycled to the reactor at moderate rates to prevent shear forces from disrupting the floc structure. This separation of solids and liquids is a crucial operation in the contact digester and the removal of gas-producing particles is difficult. Gas-stripping or cooling of the effluent en route to the separator may counteract the problem: a shock temperature reduction from 35–15 °C arrests gas production in the settler and enhances the flocculation of solids. The lat-

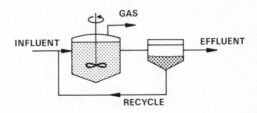

Fig. 19. The contact process

97

ter can also be achieved by the use of coagulants such as sodium hydroxide, followed by ferrous chloride. The utilisation of an ultrafiltration membrane to achieve high cell retention has also been considered [20, 21]. Lane [22] detailed problems in the separation of solids from discharged mixed liquor, because of the continuation of gas production in the settler due to high ambient temperatures. The application of a vacuum to the settling tank feed will degasify the liquid, however, and reduce the problem. Sludge concentration in the contact process rarely exceeds 5–10 g VSS l^{-1}, and the superficial liquid velocity in the settler should not be greater than about 1 m h^{-1} to allow for sufficient settling of flocculated sludge.

The anaerobic contact process was initially developed to treat meat packing waste and has subsequently been used for food-processing and other wastes. The process was one of the first of many anaerobic digestion systems to incorporate the retention of microorganisms in the digester independent of HRT. Instability in the contact process can be attributed to inactivation of the digester biomass, waste composition and also indirectly to the retention time of solids materials. The fruit and vegetable processing industries, for example, produce wastes rich in carbohydrates; these are rapidly converted to volatile acids by the acetogenic bacteria and the resulting fluctuations of pH have detrimental effects on the slower-growing methanogenic reactor population.

A pilot plant contact system designed by Lane [22] employed a tank of working capacity 23 m^3, with a floating gas holder similar to those in use with conventional CSTR systems, of capacity 10.5 m^3. The gas produced in the anaerobic digestion process was used, at a rate of 20 l min^{-1}, to agitate the digester contents by passage through a draught tube apparatus. The latter was provided with a jacket through which water was circulated at 12 l min^{-1}, to maintain the reactor temperature at 36 ± 2 °C. A galvanised iron shell containing polystyrene beads comprised the insulation. Discharge of the mixed digesting sludge from digester to settling tank occurred by gravity flow and agitation within the settler disentrained attached gas bubbles. The disadvantages of the design included the tendency of the sludge to float, leading to difficulties in solids recycle: this was the result of continuous gas evolution in the settler unit, with gas bubbles adhering to flocs and causing them to rise in the vessel. The thermophilic contact process employed by Schlegel and Kalbskopf [23] also manifested sludge flotation, but at the elevated temperatures of the system poor sedimentation was ascribed to the lack of flocculation of the digester biomass and low biomass production. Sludge settling characteristics tend to deteriorate at sludge loadings in excess of 0.25 kg COD kg^{-1} VSS d^{-1}, and biomass separation from the medium becomes more difficult above a mixed liquor VSS concentration of 18 g l^{-1}; gas production in the sedimentation vessel enhances these limitations [11]. However, gas-cooling or stripping prior to effluent discharge to the settler are effective remedies to continued gas-production, although these inhibit the activity of methanogenic bacteria.

Efficient settling may not be achieved under conditions of high influent SS in anaerobic contact reactors, although the small but significant SS concentration of bean blanching waste, for example, may be treated effectively by a contact system. High rate treatment in general is also unlikely to be efficient in this type of reactor, as influent SRT, if prolonged to any extent, will tend to lead to the displacement of biomass.

The flocculated biomass is necessarily important in the contact process. Overvigorous mixing may disperse the microbial structures of consequence in a single-stage reactor. These structures include the consortia composed of the OHPA bacteria and the hydrogen-removing methanogens. Conversely, insufficient agitation may lead to compaction of the biomass and therefore to reactor failure. The sludge flocs present in the digester have sludge volume index (SVI) values of 70–150 ml g^{-1} [24, 25]. The SVI value is dependent upon the type of wastewater and the sludge load. A high sludge load causes bulking sludge.

6.2.2 Process Efficiency

In anaerobic reactor designs where floc characteristics are important for operational performance any environmental change which tends to alter these characteristics is disadvantageous as process instability and often failure is the ultimate result. Organic overloading or large variations in loading and input of toxic substances are therefore liable to be less well tolerated in suspended floc processes than in, for example, fixed-film systems such as expanded or fluidised beds. In recycled floc reactors, a lower organic loading rate (0.25 kg COD kg^{-1} VSS d^{-1}) improves the settleablity of the sludge [26, 27] but may decrease floc formation, although higher loading rates have been tolerated with reasonable success in some contact systems [28].

The contact process has been successfully employed on a small scale for the treatment of animal wastes [29] although minimal HRTs of 12–15 days were necessary for the particular feedstock type. The biogas produced, however, could be converted through a co-generator to provide a continuous supply of electricity, while the engine cooling water was utilised to heat the digester thus making the process economical.

COD reductions of between 90 and 95% can be achieved in wastewaters with COD values in the range 2–10 g l^{-1} when high volumetric loadings are applied [11]. BOD$_5$ loadings of 0.44–2.5 kg m^{-3} d^{-1} are frequently found with HRTs of between 0.5 and 5 days, and 70–98% BOD removals have been reported [30] although lower operational efficiency in the treatment of domestic sewage was observed. The anaerobic contact process is reported generally for the treatment of high strength wastewaters, as the protracted retention times necessary for the conversion of dilute wastes renders the system impractical: secondary treatment of effluent may be necessary, and problems in sludge separation have been recorded [31]. However, a total organic carbon (TOC) removal efficiency of 90% at organic loadings of 2 kg TOC m^{-3} d^{-1} of dilute wastewater (1.0 g l^{-1}) was reported at HRTs of 3–6 h in an anaerobic contact digester, the optimum loading rate being 1.7 kg TOC m^{-3} d^{-1} [32].

Relatively high loading rates of 10 kg COD m^{-3} d^{-1} of complex wastes such as rum stillage have been applied to contact digesters; these contained little readily biodegradable carbohydrates. Low loading rates of around 2 kg COD m^{-3} d^{-1} were found to be necessary for the effective treatment of carbohydrate-rich potato wastes, and frequent reinoculation of bacteria into the system was required [28]. COD removal efficiencies of 70–90% were claimed for all the high strength wastes tested; these were found to depend upon settling efficiency but not upon the SS concentration of the waste. Low settleability caused high SS content and therefore high COD values in the effluent. Efficient process performance in the anaerobic contact system hence requires reactor biomass of high settleability, and seeding and start-up need careful control.

Effluents from starch manufacturing processes, wine distillation and yeast production, at COD values of 10, 22, and 45 g l^{-1} respectively, were treated by a reversed-flow type clarigester. Loading rates of 2.4 kg COD m^{-3} d^{-1} (24 °C), 3.2 kg COD m^{-3} d^{-1} (33 °C) and 4.0 kg COD m^{-3} d^{-1} (35 °C) respectively were applied to the system and COD reductions of 97% and 80% achieved [33]. The reactor was limited by the control of sludge return, as the solids in the digester were required to settle back into the reaction vessel under the influence of gravity. Sludge-wasting was considered unnecessary as solids-loss in the effluent was balanced by synthesis of biomass.

A high rate contact digester, which is reported capable of producing high specific yields of CH$_4$ has been demonstrated; a wide range of wastes containing low and high strength suspended solids were tolerated and HRTs of 3 days were reported at organic loading rates of 7.5 kg COD m^{-3} d^{-1} [34]. This reactor configuration was also utilised for the stable and efficient treatment of cane juice stillage at loading rates in excess of 5 kg COD m^{-3} d^{-1} [35] but an operational limit was reached due to the predominantly high sulphate levels of the stillage. Following nutrient assessment, however, continuous iron addition improved digester performance substantially, although iron sulphide precipitation increased the requirement for flocculant. This latter necessity would hence tend to decrease the economic viability of the process.

Despite the limitations of the anaerobic contact system, its advantages generally outweigh its disadvantages to such an extent that, notwithstanding the emergence of so many more sophisticated anaerobic treatment techniques, the contact process is still one of the most widely used.

6.3 The Upflow Anaerobic Sludge Blanket Reactor

6.3.1 Design and Operation

The upflow anaerobic sludge blanket (UASB) reactor (Fig. 20) was initially developed for widespread use in the Netherlands. Inert media are generally absent from the system and the biomass is maintained in suspension by gas bubbles. The bacteria develop as a flocculant mass in an upward-flowing wastestream. The microbial "blanket" is retained by its own mass and by baffles or screens forming the settler unit in the upper portion of the reaction vessel, whilst gas and liquid escape from the top of the tank. As dissociation of the bacterial mass does occur to some degree, organisms are lost in the outflow, but the mean detention time

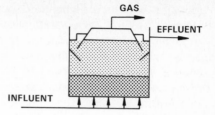

Fig. 20. The upflow anaerobic sludge blanket (UASB) reactor

of the bacteria is protracted enough to allow the growth of a dense mass of methanogenic microorganisms although liquid detention times are low.

In the UASB reactor, the biomass is present as compact grains or granules of up to 3–4 mm in diameter, which develop by some as yet undefined mechanism under the continuous upflow conditions [36, 37]. One of the two fundamental design principles for maintenance of high sludge retention in the UASB reactor is founded on sludge with improved sedimentation properties. Larger granules form the sludge bed or lower portion of the reactor, developing after a few months of system operation [38]. Sludge settleability characteristics improve if mechanical agitation of the sludge bed is minimal or absent. The sludge concentration of the bed is about 40–70 g VSS 1^{-1}, and the particles have a high settling rate of around 50 m h^{-1} [36]. Above the sludge bed, the sludge blanket develops; the blanket consists of smaller grains, flocs and gas bubbles and ranges from dense and granular particles with high settling velocities near the base to lighter, less dense grains higher in the blanket (Fig. 20).

Granule characteristics may be detrimentally affected by significant amounts of suspended matter in the influent wastestream. Granule density can be increased by inclusions of particulate inert matter resulting in settling of the sludge; sedimentation of inert particulates renders the reactor susceptible to blockage and also effectively reduces digester active volume. By virtue of their high settling velocities and activity, the pellets which develop under the continuous upflow regime in the system are retained in the UASB at high concentrations, to produce correspondingly high digestion rates of up to 50 kg m^{-3} d^{-1} [36]. Effective treatment of high strength wastes from for example the food processing industries is therefore possible.

The UASB system is mixed by hydraulic upflow and rising gas bubbles, and improved agitation can be achieved by the use of several influent ports to deliver feed to the vessel. COD removal occurs throughout the bed and blanket [38]. As the sludge concentration in the blanket is in the region of 10–30 g VSS 1^{-1}, the average reactor sludge concentration is about 20–40 g VSS 1^{-1} [36]. An investigation into the effect of sludge loading rate on granulation in UASB systems indicated that pellets formed only at loading rates in excess of 0.6 kg COD kg^{-1} VSS d^{-1}, while at 0.3 kg COD kg^{-1} VSS d^{-1} bulking and washout occurred [39].

The second main design principle of the UASB is the installation of a gas/solids separating or settling device in the upper part of the reaction vessel. The smaller particle size and flocculation characteristics of the blanket zone give rise to a settling rate inferior to that found in the bed, thus to permit retention of the blanket sludge, the applied liquid velocity in the settler unit should be relatively

101

low (0.5–1.0 m h^{-1}): higher liquid velocities tend to produce unacceptable sludge losses. These lower velocities, however, allow the accumulation of any sediment from the influent wastewater in the reactor and in the sludge, resulting in decreased sludge activity. The UASB is not, therefore, an effective treatment process for wastewater containing high suspended solids concentrations and organics in the wastestream must possess a proportionately high degree of solubility. The toal suspended solids (TSS) should ideally be limited to around 500 mg l^{-1} [38].

A model was developed and experimentally tested for the physical behaviour of sludge in the blanket zone of UASB reactors [40]. The model was based on mass balance for the sludge in an upflow reactor relative to the biogas production, settling properties and linear fluid velocity in the system. As the experimental determinations were made using wastewater containing lower fatty acids as the main organic pollutants and anaerobic sludge of good settleability, the predictive value of the model is limited, although it may have applications in the second (methanogenic) stage of anaerobic wastewater treatment.

Mixing in the UASB promotes a greater degree of contact between the wastewater and the bacterial flora, but internal or external separation of solids and liquid/gas is still necessary. Ensuring even flow-distribution and avoiding influent bypass presents problems, as does start-up. The gas/solids separator device incorporated into the UASB system of Schwartz et al. [41] consisted of a plastic ring extruding from the upper portion of the reactor for a distance of 1.5 cm and at an angle of 45 °, on top of which was placed an inverted plastic cone. The uppermost part of the cone had an opening of 3 cm through which the majority of the gas would pass into the gas space above the effluent port during reactor operation; the lower part of the cone was slatted and in contact with the reactor wall. This design is similar to that employed in many UASB systems [42–44] although variations in the form of baffled plates and sweeping and skimming arms also exist [37, 38]. The 55 ° angled wall of the cone in the reactor of Schwartz et al. [41] was designed to aid the settling of the sludge and thus enhance retention in the reaction vessel. Lettinga [43] recommended a solids settler wall angle of 50 °, which appeared to be more efficient for solids retention in the reactor. A smaller opening for gas venting would also increase the overall surface area available for solids settling and enhance efficiency, although Schwartz et al. [41] recorded adequate stability with their reactors. Excellent stability was claimed for a UASB laboratory-scale reactor in the absence of a gas/solids separator [45], the treated citrus peel press liquor overflowing from the top of the digester containing no settleable solids after the second week of operation.

The settling process allows the reactor solids to be recycled back to the system; long SRTs are hence maintained even with short HRTs and in consequence, the size of the reactor (and hence the cost) can be significantly reduced. The bacterial population of the system can also develop more quickly in a regime incorporating recycle, permitting improved stability and reducing start-up times after interruption. The quantity of biomass in the reactor can be maintained at a constant level by means of the periodic wasting of small amounts of sludge [46].

One of the most serious limitations of the sludge blanket process is the considerable time (4–8 weeks using seed sludge) involved in the initiation of digestion. The washout of sludge during the initial phases of operation is significant, and the consequent loss of net bacterial growth with the effluent engenders a depression in the retained sludge volume and stagnation in the gas production phase of the reaction. Composition of the wastewater plays a major part in the operation of UASB reactors: granulation of sludge proceeded satisfactorily in one investigation, when yeast, sugar beet and potato wastes formed the substrates, whereas problems arose with distillery, corn starch and rendering wastes [39].

Initial seeding with an active digester sludge is a prerequisite for effective start-up of a UASB reactor, as are prevention of pH and toxic shocks, gradual loading increases and extended HRTs. Start-up, therefore, is dependent upon an equilibrium between loading and washout as well as the selection of a suitable seed sludge, wastewater composition and careful system management; the addition of fine sand or other solids is not uncommon as a start-up aid. Design plays an important part in system efficiency and operational performance is strongly dependent upon effective sludge retention and activity.

6.3.2 Process Efficiency

The UASB has been reported to perform well in both the thermophilic and the mesophilic ranges [41], although pH in the thermophilic reactors employed was slightly higher than the optimum range. Substantial buffering developed within the reactors within a few weeks subsequent to start-up, and no further pH adjustment to feedstock was required. Some investigators have reported the advantages in maintaining pH at or near neutrality in UASB reactors [38, 46]; reduction in COD removal is the frequent result of pH fluctuations. Thermophilic UASB systems typically produce a gas with a CH_4 content of approximately 3–4% less than that of their mesophilic counterparts. A comparison of 35 °C and 50 °C operations in UASB reactors suggested that the systems maintained in the mesophilic range were more efficient in terms of COD and BOD removal and TSS reduction, at organic loadings of greater than 18 g COD l^{-1} d^{-1} [41]. A sucrose/glucose influent applied to UASB reactors under thermophilic conditions and at a loading rate of 45 kg COD m^{-3} d^{-1} was treated with an efficiency of 80% within three months of system initiation [47].

Increased loading is tolerated with little loss of stability by UASB reactors, but variation in COD and BOD removal with increasing organic loading is not always consistent [41] and erratic TSS removal rates have been observed [38]. High suspended solids concentrations often recorded in sludge blanket system effluents may be in part a result of the design of the gas/solids separator, or intermittent blockage of some parts of the settler apparatus. Lane [45] reported that the efficiency of a UASB reactor improved considerably as the organic load increased and postulated that this was an effect of the conditioning or adaptation of the sludge with time. At an organic loading rate of 11.15 g COD l^{-1} d^{-1} and an HRT of 7 days, 95% COD removal was achieved. The greatest washout rate (and therefore lowest sludge concentration) in a typical UASB system was observed to occur at low organic loading rates [39].

At a constant temperature of 20 °C and a waste strength of 500 mg COD l^{-1}, loadings of 0.25–0.47 kg m^{-3} d^{-1} raw sewage were applied to a laboratory scale

UASB followed by an anaerobic filter. COD removals of 78–90% were obtained [48] and a pilot plant was constructed on the basis of these results. Superior results were achieved without mixing in the reactor, an observation subsequently confirmed by Heertjes and van der Meer [32] who reported that mechanical stirring did not improve the performance of UASB systems: 90% TOC removal was reported at the optimum loading of 2.0 kg TOC m^{-3} d^{-1} of dilute wastewater (1.0 g l^{-1}) and at HRTs of 3–6 h.

One of the major problems of the UASB reactor design has been its sensitivity to feed interruptions: generally, a complete recycle of the entire system is necessary if process or instrumentation failure occurs [38, 45, 49]. Although feed interruption has a severe adverse effect on the system, shock-loading of substrate appears to be better tolerated; in a methanol-converting UASB process, shock-loads were observed to have a stimulatory effect on reactor activity [46].

A number of wastestreams from various sources are amenable to treatment by the UASB process. Skimmed milk wastewater with a COD value of 1 500 mg l^{-1} loaded at up to 7.0 kg m^{-3} d^{-1} was treated at efficiencies of 90% at 30 °C [50]. Due to the considerable progress made in the technical aspects of the UASB, the practical feasibility of the process has been proved for the treatment of wastewaters from the sugar-beet and potato processing industries [46, 51] and alcoholic waste from a chemical source [49]. COD loadings of between 4 and 14 kg m^{-3} d^{-1} of sugar-beet waste with a COD of 5 000–9 000 mg l^{-1} have been applied to a UASB system, with 65–95% COD reductions obtained [49, 50]. In a system treating acetate-rich wastes, lower removals of up to 70% were noted, and 1.6 kg COD m^{-3} d^{-1} was found to be optimum for biogas production; the system was self-scrubbing, however, and the CH_4 content of the gas reached 99% [53]. Filtered slaughterhouse wastes were treated with 80% efficiency on a pilot scale UASB at temperatures below 20 °C [54], although in general mesophilic and thermophilic operations are typical of these type of reactors.

Twelve UASB plants have been constructed in the Netherlands since 1978 for treating high strength food industry wastes, with volumes of 500–5 500 m^3 and operational loads of 10–15 kg COD d^{-1} [36]. Potato starch, potato cooking and beer wastes have been treated in reactors ranging from 13 000–66 000 kg COD d^{-1} treatment capacity, and COD reductions of 80–85% achieved [15]. The projected performance characteristics of a full-scale UASB designed for use at a food processing plant in Wisconsin, USA included a COD reduction of 80% at a loading capacity of 6 kg m^{-3} d^{-1} at 26 °C [38]. The UASB configuration has been estimated to be particularly suited to the treatment of wastes with low suspended solids and high dissolved organic substances, such as sugar wastes, although a major problem in such systems is the high ammonia content which can be deleterious to microbial pellet production and thus to full efficiency of operation [29].

References

1. Hobson PN (1982) Production of biogas from agricultural wastes. In: Subba Rao NS (ed) Advances in agricultural microbiology. Butterworth Scientific, London, p 523

2. van den Berg L (1977) Can J Microbiol 23:898
3. Lehmann V, Wellinger A (1981) Biogas production from full-scale on-farm digesters. In: Vogt F (ed) Energy conservation and use of renewable energies in the bioindustries. Pergamon, Oxford, p 353
4. Anderson GK, Donnelly T (1978) Anaerobic contact digestion for treating high strength soluble wastes. In: Mattock G (ed) New processes of wastewater treatment and recovery. Ellis Horwood, Chichester, p 75
5. Tapp MDJ (1981) A commercial biogas producing plant. In: Vogt F (ed) Energy conservation and use of renewable energies in the bioindustries. Pergamon, Oxford, p 473
6. Mills PJ (1979) Agric Wastes 1:57
7. Morris JE (1980) The digestion of crop residues – an example from the Far East. In: Stafford DA, Wheatley BI, Hughes DE (eds) First Int Symp on Anaerobic Digestion, Cardiff, 17–21 Sept 1979. Applied Science Publishers, London, p 289
8. Lin Chou W, Speece RE, Siddiqi RH (1979) Biotechnol Bioeng Symp 8:391
9. Converse JC, Graves RE, Evans GW (1977) Trans Am Soc Env Eng 1977:336
10. Van Velsen AFM (1977) Neth J Agric Sci 25:151
11. Mosey FE (1981) Trib Cebedeau 34:389
12. Kandler O, Winter J, Temper U (1981) Methane fermentation in the thermophilic range. In: Palz W, Chartier P, Hall DO (eds) Energy from biomass. Applied Science Publishers, London, p 472
13. Maurer K, Pollack H (1983) Rep Contractors Meeting of European Comm, Paria, Italy, R and D
14. De Faveri HP, Nyns EJ (1981) Slaughterhouse waste treatment by biomethanation. In: 2nd Int Symp on Anaerobic Digestion, 6–11 Sept., Travemünde, Germany
15. Klass DL (1984) Science 223:1021
16. Rimkus RR, Ryan JM, Cook EJ (1982) J Water Pollut Control Fed 54:1447
17. Therkelsen HH (1979) J Water Pollut Control Fed 51:1949
18. McConville T, Maier WJ (1979) Biotechnol Bioeng Symp 8:345
19. Callander IJ, Barford JP (1983) Biotechnol Lett 5:147
20. van den Heuvel JC, Zoetemeyer RJ, Boelhouver C (1981) Biotechnol Bioeng 23:2001
21. Sutton PM, Li A, Evans RR, Korchin S (1982) Dorr-Oliver's fixed film and suspended growth anaerobic systems for industrial wastewater treatment and energy recovery. In: Proc 37th Ind Waste Conf, Purdue University, Lafayette, Indiana 1982. Ann Arbor Science, Ann Arbor Michigan, p 667
22. Lane AG (1984) Environ Technol Lett 5:141
23. Schlegel S, Kalbskopf KH (1981) Treatment of liquors from heat treated sludge using the anaerobic contact process. In: 2nd Int Symp on Anaerobic Digestion, 6–11 Sept., Travemünde, Germany
24. Schroepfer CJ, Fullen WJ, Johnson AS, Ziemke NR, Anderson JJ (1955) Sewage Ind Wastes 27:460
25. Anderson GK, Donnelly T, Letten DJ (1980) Anaerobic treatment of high-strength industrial wastewaters. In: 3rd Int Congress on Ind Wastewater and Wastes, Stockholm
26. Donnelly T (1978) Process Biochem 13:14
27. Anderson GK, Duarte AC (1980) Environ Technol Lett 1:484
28. van den Berg L, Lentz CP (1980) Performance and stability of the anaerobic contact process as affected by waste composition, inoculation and solids retention time. In: 35th Ind Waste Conf, Purdue Univ, Lafayette, Indiana 1980. Ann Arbor Science, Ann Arbor Michigan, p 496
29. Stafford DA, Etheridge SP (1983) I Chem E Symp 77:141
30. Schroepfer GJ, Ziemke NR (1959) Sewage Ind Wastes 31:164
31. Simpson DE (1971) Water Res 5:523
32. Heertjes PM, van der Meer RR (1979) Comparison of different methods for anaerobic treatment of dilute wastewaters. In: Proc 34th Ind Waste Conf, Purdue Univ, Lafayette, Indiana 1979. Ann Arbor Science, Ann Arbor Michigan, p 790
33. Cillie GG, Henzen MR, Stander GJ, Baillie RD (1969) Water Res 3:623
34. Stafford DA (1983) Biotechnol Lett 5:639
35. Callander IJ, Barford JP (1983) Biotechnol Lett 5:755
36. Heijnen JJ Development of a high rate fluidised bed biogas reactor. In: Anaerobic Wastewater Treatment: Proc Eur Symp, Nov. 1983, Noordwijkerhaout, Netherlands, p 259
37. Ross WR (1984) Water SA 10:197

38. Christensen DR, Gerick JA, Eblen JE (1984) J Water Pollut Control Fed 56:1059
39. Hulsoff-Pol LW, de Zeeuw WJ, Velzeboer CTM, Lettinga G (1983) Water Sci Technol 15:291
40. Buijs C, Heertjes PM, van der Meer RR (1982) Biotechnol Bioeng 24:1975
41. Schwartz LJ, De Baere LA, Lanz RW (1982) Biotechnol Bioeng Symp 11:463
42. Lettinga G, van Velsen AFM, de Zeeuw W, Hobma SW (1979) Feasibility of the upflow anaerobic sludge blanket (UASB) process. In: Proc Nat Conf on Environ Eng, San Francisco, July 9–11. ASCE, p 35
43. Lettinga G (1980) Anaerobic digestion for energy saving and production. In: Int Conf on Energy and Biomass, Brighton
44. de Zeeuw W, Lettinga G (1983) Acclimation of digested sewage sludge during start-up of an upflow anaerobic sludge blanket (UASB) reactor. In: Proc 35th Ind Waste Conf, Purdue Univ, Lafayette, Indiana. Ann Arbor Science, Ann Arbor Michigan, p 39
45. Lane AG (1983) Environ Technol Lett 4:349
46. Lettinga G, van der Geest ATh, Hobma S, van der Laan J (1979) Water Res 13:725
47. Wiegant WM, Claasen JA, Borghans AJML, Lettinga G: High rate thermophilic anaerobic digestion for the generation of methane from organic wastes. In: Anaerobic Wastewater Treatment: Proc Eur Symp Nov 1983, Noorwijkerhaout, Netherlands, p 392
48. Pretorius WA (1971) Water Res 5:681
49. Lettinga G, de Zeeuw W, Ouborg E (1981) Water Res 15:171
50. Lettinga G, van Velsen AFM (1974) H_2O 7:281
51. Lettinga G, Pette K Ch, de Vletter R, Wind E (1977) H_2O 10:526
52. Lettinga G (1978) Feasibility of anaerobic digestion for the purification of industrial wastewaters. In: 4th Eur Sewage and Refuse Symp. EAS, Munich
53. Godwin SJ, Wase DAJ, Forster CF (1982) Process Biochem 17:33
54. Sayed S (1981) Anaerobic treatment of slaughterhouse waste. In: 2nd Int Symp on Anaerobic Digestion, 6–11 Sept., Travemünde, Germany

7 Single-Stage Fixed-Film Filter and Contact Processes

The settling and recycling of biomass present difficulties in the anaerobic treatment systems which depend upon freely suspended bacterial growth. Several techniques have elaborated on these processes by immobilisation of the biomass on or around carrier particles or inert surfaces. The anaerobic filter and rotating biological contactor systems require a relatively large quantity of inert media whereas the carrier-assisted contact process utilises very little.

7.1 Anaerobic Filters

7.1.1 Design and Operation

The wastewater in the anaerobic filter (fixed bed, fixed-film) system passes the reactor usually with vertical flow, either upflow or downflow (Fig. 21) (although a horizontal flow unit has been described by Landine et al. [1]). The early investigation of upflow media processes in relation to the anaerobic conversion of wastewaters demonstrated that solids removal and gasification of the waste was feasible. The anaerobic filter assembly was first assessed by Coulter et al. [2] and further developed by the work of Young and McCarty [3]; these workers illustrated in their extensive laboratory studies the potential of upflow, fixed-film anaerobic bioreactors for the treatment of dilute organic wastestreams and the production of biogas of CH_4 content of up to 75%.

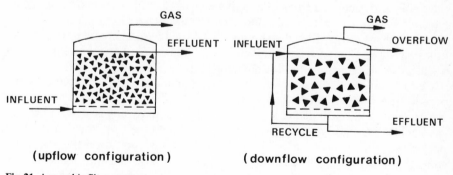

Fig. 21. Anaerobic filter reactors

The microorganisms in the anaerobic filter reactor attach to an inert medium or become entrapped. In the upflow design of reactor (Fig. 21) the waste flows upward through the support matrix and biogas is collected at the top. The system can function at ambient temperatures or be heated, or a hot wastestream may be utilised to maintain reactor temperature at a specified level. The bacterial mass in and on the inert medium rapidly degrades the substrates in the wastestream, and although the biomass tends to disentrain at intervals, its overall detention time in the digester may be of the order of 20 days, thus permitting the growth of methanogenic bacteria. Once a suitable microflora has developed to a sufficient degree it can be maintained permanently in the reactor to ensure continuous growth [4].

Various media types have been employed in both upflow and downflow anaerobic filters, of average diameter 0.2 mm–6.0 cm and of both porous and nonporous structure [5]. Downflow filters may suffer blockages if an excessively small medium is employed [6] and to minimise this, filter media tend to have relatively large diameters (>2.0 cm). The surface area of medium per m^3 of reactor is therefore limited to values of approximately 200 m^2 m^{-3}. Due to this restricted area the sludge grows in thick layers of the order of 1–4 mm around the packing particles and diffusion limitation probably occurs to some extent. The sludge concentration has been found to be between 10 and 20 g VSS l^{-1} and the filter may still be susceptible to blockages from solids [7].

The surface roughness of the packing filter media and the degree of porosity, in addition to pore size, affect the rate of colonisation by bacteria. A variety of materials have been examined as supporting matrices for use in downflow stationary fixed-film reactors by van den Berg and Kennedy [8]. These included polyvinyl chloride sheets, potters' clay, red drain tile clay, needle-punched polyester and glass. The red drain tile clay proved to be the most efficient support in terms of start-up and steady-state reactor performance, slightly superior to the needle-punched polyester. Reticulated polyurethane foam was considered an excellent colonisation matrix [5] although Fynn and Whitmore [9] reported that the weak binding forces between the methanogens and such particles could impose a limit on process intensification in fixed bed methane digesters. Media used in full-scale and pilot tests conducted to date have ranged from rock to a number of commercially available plastic and ceramic tower packings [10], in addition to a variety of material such as cloth and brick [11], gravel [12], crushed stone [13], Raschig rings [14], anthracite, and granular activated carbon [15]. The relation of media type to waste has not been widely documented although it has been indicated that small diameter, loose-fill media should not be used with high strength carbohydrate wastes because of the rapid increase in biomass synthesis with this type of waste and hence the potential for plugging [11].

A substantial percentage of the biomass of an anaerobic filter occurs as suspended flocs trapped in the spaces between the media particles. In general, a recycle facility is not included in the system design i.e. plug flow is maintained, but gas bubbles produced during reactor operation tend to stir up flow. However, a plug-flow regime may cause build-up of toxic substances or lower the pH, hence recycle may be employed to counteract these problems. Entrapped gas bubbles can cause clogging: filter backwashing or draining-off of the suspended solids be-

108

tween inert media particles will alleviate this [16]. The diffusely-attached biofilm is responsible for a significant part of the biological activity in fixed bed reactors but the non-attached biomass plays the major role in anaerobic conversion. The development of an active biofilm is promoted by porous inert supporting media and backwash, bubble-generation and hydraulic shear may control biofilm thickness.

In the upflow reactor design the biomass accumulates mainly as microbial flocs whereas in downflow systems the biomass is retained almost entirely as a film on the reactor walls and media. The treatment of soluble organic wastes has predominated in research into the use of anaerobic filter systems although selection of suitable media, in addition to efficient system design and operational principles, should permit the conversion of wastestreams with significant fractions of suspended solids. If the suspended solids are biodegradable and evenly dispersed the direct application of waste to the filter should produce few adverse effects. Detailed analyses have shown that a major fraction of the suspended solids in fixed bed reactors are held loosely within the interstitial spaces of the matrix and about one quarter to one half of the total mass is held by attachment to the media surfaces [17].

Waste treatment in anaerobic packed bed reactors is frequently initiated by seeding the system with active bacteria from an operational steady-state anaerobic digester. The maturation process ranges from 3–10 weeks and comprises the development of the microorganisms responsible for waste conversion and CH_4 genesis, either attached to the media surfaces or accumulated in the interstitial void spaces by a combination of flocculation and settling. Constant effluent quality and gas production rate indicate steady-state.

One particular reactor design incorporated an underflow distribution manifold and a media support grating [17]. The media, of modular cylindrical corrugated blocks, provided a surface area of $100 \text{ m}^2 \text{ m}^{-3}$ per unit volume and a porosity value in excess of 95%. Shortcircuiting along the column wall was minimised by dispersion rings between each cylindrical medium section. The laboratory-scale anaerobic filters of Schwartz et al. [18] employed quartzite rock of 2.5–4.5 cm in diameter supported by diffusion plates above a mixing chamber for the influent feedstock. The columns had porosity values of 50%; the reactors were operated at mesophilic (35 °C) and thermophilic (50 °C) temperatures. Both designs included gas/effluent separator devices and neither experienced blockage problems.

According to Young and Dahab [11], the plug-flow mode of operation of anaerobic upflow filters results in high rates of bacterial growth and waste removal at lower reactor heights. Steady-state COD profiles throughout the active height typically manifested little additional removal of soluble COD above 1 m in the reactors employed and the investigators concluded that for equivalent volumetric loadings, shallow reactors tended to provide more efficient treatment than taller units. However, reactor heights and media depths of less than 2 m may not prove efficient because of the risk of excessive solids washout; in addition, the conditioning of solids in the upper reactor zones causes reduced solids activity via biological decay and increased retention in the reaction vessel due to solids flocculation. Suspended solids in shallow anaerobic filters have also been demonstrated to decrease the quality of effluent produced to a significant extent [14]. In the 1.83 m reactor of Dahab and Young [17] the majority of COD removal occurred in the first 0.3 m of column height at a loading rate of 0.5 g COD l^{-1} d^{-1}, but biological activity progressed up the column with increasing loading rate.

Upflow anaerobic filter performance relates strongly to media composition and structure [16]. It is probable that shape and void size of the media are more important than actual surface area for the establishment of satisfactory treatment performance: improved COD removal was achieved using plastic modular media of unit surface area 98 m^2 m^{-3} than with similar smaller media of unit surface area 138 m^2 m^{-3}.

7.1.2 Process Efficiency

Descriptions of the operation and performance of up- and downflow fixed-film bioreactors have been provided by several investigators. The "Anflow" system was developed as an energy-conserving wastewater treatment and a demonstration pilot plant, designed in the upflow mode to treat low strength, low temperature wastestreams has been reported [19]. At temperatures in the range 10–25 °C, efficient treatment of dilute sewage wastes was recorded and substantial operational tolerance of temperature effects in the anticipated working ranges noted. The fully-operational pilot plant succeeding the feasibility study was subjected to temperature changes from 10–25 °C: no effects on BOD removal rates were recorded. BOD levels did not meet the required standards under every circumstance, however, and a final polishing treatment was necessary [19]. The bioreactor could operate for extended periods without plugging and off-gases produced and recovered contained 60–70% CH_4 by volume. The anaerobic filters of Schwartz et al. [18], using a feed of thermally conditioned sludge decant liquors, were reported to perform well at both mesophilic and thermophilic temperatures; at 0.15 kg COD m^{-3} d^{-1} organic loadings a COD removal in excess of 55% was achieved. Changes in organic load did not produce instability and recovery after months of down time was stated to be excellent. Methane was produced at steady state in the range 0.33–0.43 m^3 kg^{-1} COD removed and the total gas evolved contained 60–68% CH_4. The mesophilic reactor did not suffer blockage, although the quartzite media were coated in a substantial layer of solids. The mesophilic digester was found to be superior in total solids reduction and BOD and COD removal except at organic loadings above 0.2 kg m^{-3} d^{-1}, where thermophilic treatment proved more effective, observations in agreement with those of Basu and Leclerc [20].

The temperature range of the downflow anaerobic filter is generally considered to be from 10–60 °C although most investigations have been performed at 35 °C. Messing [21] found no advantage in operating a thermophilic (55 °C) over a mesophilic system and Duff and Kennedy [22] recorded instability of thermophilic downflow reactors at 30–40 kg COD m^{-3} d^{-1}, whether hydraulically or organically overloaded. Stafford and Etheridge [23] have indicated that reactors operating with extended SRT can be maintained at lower temperatures but strict temperature controls are paramount to system efficiency.

Anaerobic filters are most sensitive to changes in pH during start-up, although once steady state is achieved these reactors are relatively resistant to the effects of moderate pH changes and rapid recovery has been observed in systems exposed to pH 5.4 for 12 h [24]. pH spikes in the influent of the pilot scale filter of Genung et al. [19] were prolonged for up to 8 h at values as low as pH 3.0 and no effect on the effluent pH was observed. Considerably variable influent TSS levels (>210 mg$\,$l^{-1}) did not affect the constant effluent TSS levels in this reactor and an approximate BOD removal of 55% was claimed for the unit, with gas production reaching 100 m^3 d^{-1} at a maximum CH$_4$ content of 72%. However, flow channelling, a frequent problem in anaerobic filter systems, was reported to occur in the column, causing significantly adverse effects at low flow rates; this situation was attributed to solids accumulation in the reactor.

Waste types treated in upflow filter systems include domestic sewage, molasses distillery, pharmaceutical, vegetable processing, abattoir and synthetic stillage [12, 16, 25–29]. The treatment of both strong and weak organic effluents has been demonstrated [30] with up to 85% COD removal, while 90–99% COD removal from dairy wastes was achieved using anaerobic filters [23]. Using a synthetic filter medium of high surface area Kobayashi et al. [31] investigated the treatment of low strength domestic wastewater at 288 mg COD l^{-1}. The average BOD removal rate was 79% and the COD removal 73%. Removal efficiencies showed little sensitivity to daily fluctuations in influent wastewater quality. Filter performances at 25 °C and 35 °C were not significantly different but BOD and TSS removal declined at 20 °C. The biogas produced comprised 30% N$_2$, 65% CH$_4$, and 5% CO$_2$; ammonia, nitrogen and sulphides all increased during treatment and post-treatment was necessary. Despite this, it was considered that the system was promising for low strength wastewater treatment.

Downflow anaerobic filters have been employed to treat concentrated wastes up to 130 g COD l^{-1} [21]. Treatment of petrochemical wastestreams, with COD reductions of 93–95% have also been documented [32]. Optimum retention time was 2.3 days at a loading rate of 4.7 kg COD m^{-3} d^{-1} with 0.88 m^3 m^{-3} d^{-1} biogas produced, over 90% of which was CH$_4$. High protein fish wastes were treated in a downflow fixed-film system with up to 90% COD removal efficiencies and the loading rates applied exceeded 10 kg COD m^{-3} d^{-1} at several influent concentrations [32].

The effective performance of the pilot scale anaerobic filter of Genung et al. [19] was reported to increase with increase of COD loading rates in the range examined; at a COD loading rate of 1.2 kg COD m^{-3} d^{-1} about 1.0 kg COD m^{-3} d^{-1} was removed in the first 9 months of plant operation at an HRT of approximately 1 day. Chian and DeWalle [10] found that COD removal in a stationary fixed-film system deteriorated from 95% to 52% when the HRT was less than 7.5 days; the wastewater treated in the reactor was acidic, however, with a pH of 5.4 and a COD of 54 g l^{-1}. At a maximum loading of 0.8 kg TOC m^{-3} d^{-1} the upflow anaerobic filter of Heertjes and van der Meer [34] removed over 90% TOC of dilute wastewater (1.0 g l^{-1}) at retention times of 3–6 h; the optimum loading rate was found to be 0.5 kg TOC m^{-3} d^{-1} although CH$_4$ production peaked at loadings of around 1 kg m^{-3} d^{-1} and decreased thereafter.

Contrary to the results of Genung et al. [19], the COD removal efficiency of reactors treating alcohol stillage was reported to vary approximately inversely with organic load applied [35]. At 0.005 kg COD $m^{-3} d^{-1}$ the average removal efficiency was of the order of 84%, compared with 75% and 74% removal efficiencies at 0.01 and 0.02 kg COD $m^{-3} d^{-1}$ loading rates respectively. VFA concentration profiles indicated moreover that at progressively greater loading rates (coupled with decreased HRTs) column height became a factor of increasing importance. Total VFA effluent concentration accounted for the main part of the COD content, indicating that all components of the organic feedstock were utilised in anaerobic degradation. The treatment of high strength wastes of COD values in excess of $5 g l^{-1}$ has been reported [36]. The recycle of a portion of the effluent was used to dilute the influent stream, rendering pH fluctuations and surges of toxic substances tolerable to the system. The reactor was a full-scale unit and during one year operation was reported to be a net energy producer, the gas generated exceeding $10\,600 \times 10^6$ BTU.

The maximum loading rates applied to anaerobic filter reactors are probably capable of some degree of increase by increasing the rates of effluent recirculation. In downflow configuration systems high rate CH_4 production was found to be independent of the degree of recycle of concentrated wastes in the range 4–140 g COD l^{-1}, but recycling was beneficial when pear peeling and piggery wastes containing easily settleable SS were applied [37, 38]. Recycle rates in the downflow filter are generally 0–4 times the feed rate. Van den Berg and Lentz [39], in a comparison between upflow and downflow stationary fixed-film reactors, observed that the differences in performance of the two types were negligible, both achieving COD removals of the order of 90% at 0.5–2.6 days HRT and at loadings of 3–15 kg VS $m^{-3} d^{-1}$ of bean blanching wastes. However, basic operational parameters were reported to differ, the downflow systems acting as fixed-film processes whilst the upflow reactors appeared to operate as partially expanded or fluidised bed systems.

The severe organic overloading of downflow reactors at 10–35 °C, using bean blanching and chemical wastes and liquor from heat treated sewage digester sludge, was investigated [38]. Organic overloading to 94 kg COD $m^{-3} d^{-1}$ for a 24 h period resulted in a lag of 12–48 h prior to resumption of normal operation after overload. Loading rates and CH_4 generation rates were found to be temperature and waste dependent. Efficiencies of COD removal were little affected by temperature but were governed by waste composition, and minimum HRTs of less than one day were possible at 35 °C; HRTs were less than 3 days at 10 °C. The dilute nature of the waste ensured that the change in pH was less than one unit. At the onset of overload, the CH_4 production rate increased to a new steady state level, although not linearly; a drop in the CH_4 content of digester gas during periods of overloading has been reported elsewhere [40].

Anaerobic filters can resist high shock loads although their loading capacities are less than other fixed-film systems such as the expanded and fluidised bed designs, as a consequence of the larger-sized media required for minimisation of clogging and short-circuiting. The buffering capacity of the wastestream should, moreover, be sufficiently high to counteract drastic reductions in pH which may cause system failure. The relatively rapid responses observed to increased loading

112

rates suggest amenability to intermittent operation; this has been verified by Young and Dahab [11] and others: periods without feed ranging from days to months have been followed by return to efficient treatment capacity in a few weeks [41].

Large shock loadings of toxic materials are frequently accommodated by anaerobic filters, primarily because of the biofilm nature of the microflora; acclimation to or recovery from exposure to compounds such as chloroform and formaldehyde has been documented [42].

7.2 Rotating Biological Contactors

7.2.1 Design and Operation

Also known as the rotating disc or moving bed reactor, the RBC was described in 1928 by the filing of the original patent by A. T. Maltby [43]. However it was only with the emergence of plastics as effective and inexpensive lightweight supporting media that the aerobic process became widespread in the treatment of wastewaters; the anaerobic system is less widely documented, but potentially superior in a number of respects, including reactor size and loading rates. In the RBC, microorganisms attach to the inert plastic medium to form a biofilm; the support medium in a disc-array configuration is partly or fully submerged and rotates slowly on a horizontal axis in a vessel through which the wastestream flows (Fig. 22). As the medium rotates, the microbial film is thus exposed to the nutrients in the feedstock. The velocity of revolution through the wastewater provides some control over biofilm thickness. The discs are rotated at rates of about 1–7 rpm, frequently utilising a mechanical drive system. Generally, a plug-flow pattern predominates and excess sludge leaves the reactor with the treated wastewater. The media of anaerobic RBCs tend to be almost entirely submerged in the liquid phase and the rotating disc-array within the reactor vessel fits extremely close to the vessel walls. The liquid flow through the unit, combined with medium rotation, induces a high hydraulic shear on the biofilm. This force enhances mass transfer from substrate to microbial film and aids the sloughing-off of the biofilm [43]. Bubble generation may also contribute to exfoliation and act as a control on biofilm thickness.

There are several configurations of RBC support medium; these include discs, lattice construction and wire mesh containers of random plastic media [44]. The disc-design consists of a horizontal shaft mounted upon which, at distances of approximately 20 mm apart, are discs of 2–3 m in diameter and of the order of 10–

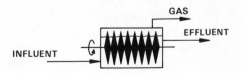

Fig. 22. The rotating biological contactor

20 mm thick. The discs may be composed of expanded or corrugated plastic material, or expanded metal. As with other anaerobic reactor configurations containing supportive media, porosity will affect biofilm formation and operational efficiency although no literature appears to deal with this aspect of RBC processes.

The anaerobic pilot-scale RBC reactors of one investigation each comprised a series of discs of inside diameter 13.97 cm and disc diameter 12.70 cm, mounted on a horizontal shaft of diameter 1.27 cm [45]. The latter was supported by external end-bearings and rotated by attachment through a pulley and belt arrangement with a variable speed drive motor. Groups of discs or stages were separated by baffles to prevent short-circuiting and the liquid flow passed through perforations in the baffles from stage to stage. Each four-staged RBC reactor was constructed of polymethylacrylate and each stage contained ten 0.318 cm thick discs spaced about 0.915 cm apart, providing a total disc surface area of 1.013 m². The first stage of each reactor was preceeded by a small mixing chamber containing an impeller which distributed the influent feed evenly through the perforations in the baffle-plate separating the mixing chamber from the primary stage. The reactors were operated with 70% of the disc area submerged, at a temperature of 35 °C.

Each stage of an RBC of the above design acts as a completely mixed reactor such that there are hydraulic and reaction discontinuities between adjacent stages. Within each stage therefore, mixing ensures even distribution of nutrients, prevents clogging and facilitates product gas transfer to the anoxic atmosphere above the liquid phase surface. The biomass mode of attachment provides adequate cell retention in the reactor and promotes the development of long mean cell residence times. However, an evaluation by McCarty [46] indicated that rotation of the disc-array was not fundamental to process efficiency and the anaerobic baffled reactor was consequently developed. The wastestream in this assembly flows over and under baffle plates and the suspended biomass is capable of vertical motion although horizontal movement through the system is restricted. Floating solids present few problems as a large liquid/gas interface exists within the reactor.

7.2.2 Process Efficiency

The RBC reactor was evaluated in laboratory scale investigations by Tait and Friedman [45]; a synthetic wastewater containing glucose as the only carbon source was utilised. Part of the glucose was replaced by methanol during the start-up period to enhance the growth of methanogenic bacteria. Volatile acid production was found to be significant in the first stage of the four-staged array. TOC removal of 80% occurred here with further reductions in the succeeding stages, each of which comprised ten 12.7 cm diameter discs; pH and alkalinity increased downstream. Increase of organic loading rate (and decrease of HRT) reduced overall soluble TOC removal from 96–76%. The degradation process was stated to be inhibited by high concentrations of the soluble substrate in the first three stages and increase of flow rate for constant conditions resulted in higher TOC

114

concentrations at most points within the reactor. Up to a loading of 21.7 g TOC $m^{-2} d^{-1}$ soluble TOC removal appeared independent of both influent concentration and flow rate. Methane and CO_2 were present in about equal proportions in the gas produced (1.76 m^3 kg^{-1} TOC removal). On the basis of the investigations carried out, a system of simple predictive equations that enabled design calculations to be made was produced [45].

The process efficiency of the baffled reactor, similar in concept to the RBC, does not appear to have been fully established, but at COD loadings of 10–20 kg $m^{-3} d^{-1}$ COD reductions of 60–80% have been recorded [47].

The large volume of biomass capable of attachment to the inert fixed medium permits accommodation of hydraulic overloading in the RBC system and the response to toxic influent should also be efficient. The treatment of high strength wastewaters is therefore feasible and practical in terms of both plant size and economy, as a ground area of only 10% of that of an equivalent conventional reactor is necessary for an aerobic RBC [44] and an anaerobic system will reduce the requirement still further. An anaerobic RBC can provide effective treatment of readily biodegradable wastewaters having COD concentrations of over 8.5 g l^{-1} [45] although upper limits of treatability have not as yet been defined. The horizontal flow through the system reduces the energy requirements, a substantial consideration in other reactors of vertical assembly and the potential for energy recovery in the form of CH_4 exists.

The HRT of RBC reactors is about 7 min for a stage comprising a fifty disc array [44] and standard loading rates are reported to be in the range 6–20 g BOD $m^{-3} d^{-1}$ for aerobic systems [48–50], although the degree of treatment depends on the waste type and system. The advantage of the RBC configuration in waste conversion processes include elimination of the channelling to which conventional percolating filters are susceptible, low sludge production and short retention times [43]. If the RBC is part of a phased anaerobic process, the incorporation of a recycle facility in the methanogenic reactor may be necessary to overcome pH variations. In addition, the shaft bearings and mechanical drive units require frequent maintenance, although improved design has resolved many early problems [50].

7.3 Carrier-Assisted Contact Reactors

7.3.1 Design and Operation

The carrier-assisted contact process is in essence identical to the contact system but with the incorporation of inert media into the reactor vessel (see Fig. 23). The addition of supportive material is extremely limited in comparison to the quantity of media commonly utilised in, for example, fluidised bed reactors and the system has been named the carrier-assisted sludge bed reactor or CASBER [51]. Small inert particles of dimensions 5–25 µm diameter are used; these have a low settling velocity and can thus be maintained in suspension with a relatively low degree of mixing. The bacteria in the system attach to the support particles, which may be of sand, anthracite or iron [52] and a substantial percentage of the active biomass exists as suspended flocs. The reactor bed is maintained in suspension by mechanical stirring, gas sparging or recycle, or a combination of these. The treated wastewater is separated from the bed in a settler assembly which is generally external to the reaction vessel; a sedimentation tank may be utilised for this purpose. The bed, in addition to the biofilm-coated particles and the suspended flocs, is recycled at moderate recycle rates [52]. As is the case in some other anaerobic processes, this separation is the most important part of the digester cycle and separation of gas-producing particles is difficult as they tend to rise to the top of the settling unit and wash out. Gas cooling or stripping can reduce the problem considerably.

In one test system the sludge present in the CASBER considered of small (0.5–1.0 mm) homogeneous flocs with fluffy exteriors and more dense interiors [51]. Significant adhesion of biomass to particles was not evident although 2–5 carrier particles were entrapped within each floc. The above investigation indicated that the major effect of the supportive media in the reactor system was the provision of increased density of sludge flocs. The complete mechanism of floc generation is not known but mechanical agitation, in combination with the gravitational effects and hydraulic and other forces in operation were probably conducive to the biomass configuration developed. The system employed in the above investigation had a relatively large number of protozoa between the flocs. Flocculation therefore, may have been partly an induced defensive response against predation, as protozoa tend to prey on free-living rather than colonial microorganisms.

The attachment of the biofloc to the support medium is a loose association; the floc thickness may therefore be primarily governed by hydraulic shear forces and bubble generation. Over-intense mechanical mixing can disperse symbiotic particles; the non-attached biomass is necessarily of great importance to system effectiveness and control of agitation is hence paramount. Eddy currents may in addition cause washout of the free-swimming protozoa.

The CASBER of Martensson and Frostell [51] consisted of a reaction tank of 35 cm in diameter and 42 cm in height, which contained 38.5 l of liquid. The agitator was mounted from the top and extended 50 mm below the liquid level in the vessel. Carrier material (5–25 μm diameter) was added to the reactor at 3% $(v \cdot v^{-1})$ prior to seeding with anaerobic sludge from a digester treating synthetic molasses waste. Temperature was maintained at 35–37 °C by a thermostatically-controlled water bath although the height of the bath (less than half the height of the reactor contents) resulted in a small temperature gradient in the reaction vessel. The effluent was led from the tank to a conical settler from which solids were recycled, initially intermittently, latterly continuously, by pump. The biogas from the CASBER was led via a condensation trap to a gas holder.

The start-up of the process was hampered by low pH values and low gas recovery and later by foaming problems. It was concluded that 25–30 days were necessary for acclimation of the system to the synthetic molasses wastewater employed. Molasses contains materials that tend to be resistant to anaerobic breakdown, as well as easily biodegradable organic matter and a bacterial adaptation period is evidently required for this and similar wastestreams.

116

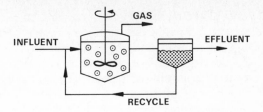

Fig. 23. The carrier-assisted contact reactor

7.3.2 Process Efficiency

The biomass concentration in a CASBER system can be maintained at around 5–15 kg SS m^{-3} [52]. Using synthetic molasses wastes at organic loadings of 3.9 kg COD m^{-3} d^{-1} and an HRT of 2.4 days, total COD reductions of up to 90% were reported [51]. The total biogas evolved was 2.03 m^3 m^{-3} d^{-1}, of which 60% was CH_4. Organic loads of up to 5.0 kg COD m^{-3} d^{-1} were treated successfully, but a load of 5.8 kg m^{-3} d^{-1} was applied without success. Bed expansion and loss of suspended solids were frequent occurrences.

When easily-hydrolysable sugar industry wastes were applied to a CASBER unit, very high organic loads of up to 24.3 kg COD m^{-3} d^{-1} at HRTs of 4.6 h were tolerated [52]. Gas production was reported to increase with increasing organic load, and total COD reductions of up to 89% recorded.

Overloading of the CASBER system with relatively complex waste substrates results in operational instability, a situation manifested by a drop in reactor pH and a decrease in biogas evolution. Careful process management is therefore a necessity if the system is to operate at maximal efficiency. The contact process (see Chap. 5) may be converted to a type of CASBER system by the simple expedient of adding inert, small diameter particles; non-biological components of some influent wastestreams, such as mud solids from sugar beet processing, can have this effect. The problems affecting both contact and sludge blanket reactors apply to a certain extent in the case of carrier assisted systems where the fraction of inert material is very low. High SS concentrations may reduce biomass activity or cause bulking of sludge by increased floc settleability; increase of the mixing rate to counteract such a problem would probably lead to unacceptable biomass losses.

The CASBER process seems to be eminently suitable for high loadings of easily-hydrolysable wastes such as those typical of the sugar industry, but limitations on wastestream composition evidently apply. The low volume of carrier media in this reactor configuration however, aids biomass retention and reduces the necessary HRT and reactor volume needed for wastewater treatment. The CASBER process in addition is suitable for the conversion of wastes which comprise non-biodegradable SS fractions as the particulate matter can settle out with the flocs.

7.4 Hybrid Reactors

An anaerobic reactor combining some properties of both the sludge blanket digester and the upflow anaerobic filter configurations has been demonstrated [53]. The packing of the upflow blanket filter (UBF) comprised plastic packing rings floating against a screen in the top third of the reactor, a 4.25 l cylindrical fermenter. A recycle facility operating at 5.4 times the feed rate was also incorporated into the design and flocculant active sludge, in combination with granular sludge from a UASB reactor was used to start-up the hybrid device. A 35 day acclimation period at 27 °C preceeded testing of operational characteristics. COD removal was found to depend upon loading rate, HRT and mixed liquor VSS. At an organic loading rate of 0.05–0.25 kg COD m^{-3} d^{-1} with an SRT of 7 days, 96% COD removal efficiency was achieved. Most of the biomass consisted of granular sludge, although filamentous forms occurred on and near the packing surface. The elevated filter section retained biomass independently of the bed. The reactor was claimed to compare favourably with other high-rate anaerobic processes in the conversion of low strength soluble wastes, due to the high biomass accumulation possible on the plastic packing.

The above hybrid system is comparable in terms of efficiency to many upflow filter and sludge blanket processes, combining the favourable points of both types. However, the start-up difficulties of such systems are probably similar to those inherent in sludge blanket reactors and may be greater. The hybrid design is more complex than those of both the filter and the blanket configurations and hence may be more liable to instability. The operational advantages must be more clearly defined if further comparisons are to be made.

References

1. Landine RC, Viraraghaven T, Cocci AA, Brown GJ, Lin KC (1982) J Water Pollut Control Fed 54:103
2. Coulter JB, Soneda S, Ettinger MB (1957) Sewage Ind Wastes 29:468
3. Young JC, McCarty PL (1969) J Water Pollut Control Fed 41:R160
4. Hobson PN (1982) Production of biogas from agricultural wastes. In: Subba Rao NS (ed) Advances in agricultural microbiology. Butterworth Scientific, London, p 523
5. Huysman P, van Meenen P, van Assche P, Verstraete W (1983) Biotechnol Lett 5:643
6. Cooper PF, Wheeldon DHV (1981) Complete treatment of sewage in a two-staged fluidised bed system. In: Cooper PF, Atkinson B (eds) Biological fluidised bed treatment of water and wastewater. Ellis Horwood, Chichester, p 121
7. Heijnen JJ (1983) Development of a high rate fluidised bed biogas reactor. In: Anaerobic wastewater treatment: Proc European Symp, Nov 1983. Noordwijkerhaout, Netherlands, p 259
8. van den Berg L, Kennedy KJ (1981) Biotechnol Lett 3:165
9. Fynn GH, Whitmore TN (1984) Biotechnol Lett 6:81
10. Chian ESK, De Walle FB (1977) Water Res 11:295
11. Young JC, Dahab MF (1982) Biotechnol Bioeng Symp 12:303
12. Sachs EF, Jennett JC, Rand MC (1982) J Env Eng Div ASCE 108:297
13. Anderson GK, Ibrahim AB (1978) Prog Water Technol 10:237

14. Plummer AH, Malina JF, Eckenfelder WW (1968) Stabilization of low carbohydrate waste by an anaerobic submerged filter. In: Proc 23rd Ind Waste Conf, Purdue Univ, Lafayette, Indiana. Ann Arbor Science, Ann Arbor Michigan
15. Khan KA, Suidan MT, Cross WH (1982) J Env Eng Div ASCE 108:269
16. Young JC, Dahab MF (1983) Water Technol 15:369
17. Dahab MF, Young JC (1982) Retention and distribution of biological solids in fixed film anaerobic filters. In: Proc Ist Int Conf on Fixed Film Biological Processes, Kings Is. Ohio, April 1982
18. Schwartz LJ, De Baere LA, Lanz RW (1982) Biotechnol Bioeng Symp 11:463
19. Genung RK, Million DL, Hancher CW, Pitt WW Jr (1979) Biotechnol Bioeng Symp 8:329
20. Basu AK, LeClerc E (1975) Wat Res 9:103
21. Messing RA (1983) Bioenergy production and pollution control with immobilized microbes. In: Tsao GT (ed) Ann reports on fermentation processes, vol 6. Academic Press, New York London, p 23
22. Duff SJB, Kennedy KJ (1982) Biotechnol Lett 4:815
23. Stafford DA, Etheridge SP (1983) I Chem E Symp 77:141
24. Clark RH, Speece RE (1971) The pH tolerance of anaerobic digestion. In: Proc 5th Int Wat Pollut Res Conf Pergamon, New York, p II/27
25. Perkins CW, Tsugita LD, Tekippe RJ (1975) Source control and biological treatment of an artichoke process plant. In: Proc 30th Ind Waste Conf, Purdue Univ, Lafayette, Indiana 1975. Ann Arbor Science, Ann Arbor Michigan, p 192
26. Arora HC, Routh T (1981) Indian Assoc Wat Pollut Control Tech Annual VI and VII:67
27. Riera FS, Valz-Gianinet S, Gallieri D, Sineriz F (1982) Biotechnol Lett 4:127
28. Braun R, Huss S (1982) Process Biochem 17:25
29. Genung RK, Hancher CW, Rivera AL, Harris MT (1982) Biotechnol Bioeng Symp 12:365
30. Newell PJ (1981) The use of a high rate contact reactor for energy production and waste treatment from intensive livestock units. In: Vogt F (ed) Energy conservation and use of renewable energies in the bio-industries. Pergamon, Oxford, p 395
31. Kobayashi HA, Stenstrom MK, Mah RA (1983) Water Res 17:903
32. Britz TJ, Botes PJ, Meyer LC (1983) Biotechnol Lett 5:113
33. Duff SJB, Kennedy KJ (1982) Biotechnol Lett 4:821
34. Heertjes PM, van der Meer RR (1979) Comparison of different methods for anaerobic treatment of dilute wastewaters. In: Proc 34th Ind Waste Conf, Purdue Univ, Lafayette, Indiana 1979. Ann Arbor Science, Ann Arbor Michigan, p 790
35. Dahab MF, Young JC (1982) Biotechnol Bioeng Symp 11:381
36. Witt ER, Humphrey WJ, Roberts TE (1979) Full-scale anaerobic filter treats high strength wastes. In: Proc 34th Ind Waste Conf, Purdue Univ, Lafayette, Indiana 1979. Ann Arbor Science, Ann Arbor Michigan, p 229
37. van den Berg L, Kennedy KJ, Hamoda MF (1981) Effect of type of waste on performance of anaerobic fixed film and upflow sludge bed reactors. In: Proc 36th Ind Waste Conf, Purdue Univ, Lafayette, Indiana 1981. Ann Arbor Science, Ann Arbor Michigan, p 686
38. Kennedy KJ, van den Berg L (1982) Water Res 16:1391
39. van den Berg L, Lentz CP (1979) Comparison between up and downflow anaerobic fixed film reactors of varying surface-to-volume ratios for the treatment of bean-blanching wastes. In: Proc 34th Ind Waste Conf, Purdue Univ, Lafayette, Indiana 1979. Ann Arbor Science, Ann Arbor Michigan, p 319
40. Graef SP, Andrews JF (1975) J Water Pollut Control Fed 46:666
41. Taylor DW (1972) Full-scale anaerobic trickling filter evaluation. In: Proc 3rd Nat Symp on Food Processing Wastes, Report no. EPA-R2-72-018. US Environmental Protection Agency, Washington, DC
42. Parkin GF, Speece RE, Yang CHJ (1981) A comparison of the response of methanogens to toxicants: anaerobic filter vs. suspended growth systems. In: Proc of the Seminar/Workshop on Anaerobic Filters. Argonne Nat Lab, Argonne, Illinois, p 37
43. Borchardt JA (1971) Biological waste treatment using rotating discs. In: Canale RP (ed) Biological waste treatment. Wiley-Interscience, New York, p 1
44. Winkler M (1981) Biological treatment of wastewater. Ellis Horwood, Chichester, p 211
45. Tait SJ, Friedman AA (1980) J Water Pollut Control Fed 52:2257

46. McCarty PL (1981) One hundred years of anaerobic treatment. In: 2nd Int Symp on Anaerobic Digestion, 6–11 Sept. Travemünde, Germany
47. Josephson J (1982) Environ Sci Technol 16:380
48. Hao O, Hendricks GF (1975) Water Sew Wks 122:70
49. Hao O, Hendricks GF (1975) Water Sew Wks 122:48
50. Benjes HH (1980) Handbook of biological wastewater treatment. Garland STPM Press, New York London
51. Martensson L, Frostell B (1983) Water Sci Technol 15:233
52. Henze M, Harremoes P (1983) Water Sci Technol 15:1
53. Guoit SR, van den Berg L (1984) Biotechnol Lett 6:161

8 Single-Stage Fixed-Film Expanded Processes

The effective and reliable treatment of industrial and other wastestreams by biological means necessarily requires a system which demonstrates wide-ranging tolerance of fluctuations in operational conditions. Temperature, influent COD and pH increases and decreases will all, to varying degrees, affect the activity and viability of anaerobic bacterial populations within the reactor systems. The CH_4-producing bacteria for example, are extremely intolerant of even minor fluctuations in environmental conditions. Immediate cessation of gas formation results from the application of a sharp temperature decrease of around 20 °C to the microbial system and this technique is in fact used in other anaerobic digester processes to aid biomass retention in settler units (see Chaps. 6 and 7). Fixed-film processes are in general considered more resistant than alternative systems to hydraulic and other shocks because of their capacity for biomass retention within a film with certain protective characteristics.

The anaerobic conversion of wastewaters is limited in suspended growth systems by sludge activity and the sludge concentration although these problems can be attenuated to some extent by settling, manipulation of wastewater composition and recycle. Fixed-film stationary processes depend on immobilised biomass; to minimise pressure losses and clogging by solids, the inert media have large diameters which in turn limit the surface area per m^3 of reactor to maximum values of around 200 m^2 m^{-3}. The biomass grows in layers of 1–4 mm in thickness around the packing material and diffusion limitations are likely to exist [1]. These problems are effectively overcome in the anaerobic expanded and fluidised bed systems (see Fig. 24) where the smaller diameter media employed enable very high biomass concentrations to accumulate as thin films around the carrier particles. Expansion or fluidisation of the media reduces or eliminates the problems of blockage, and retention times necessary for treatment can be markedly reduced.

The distinction between expansion and fluidisation in the description of anaerobic fixed-biofilm processes is not clearly defined; such considerations as me-

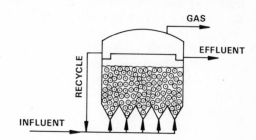

Fig. 24. The expanded/fluidised bed reactor

121

dia size and liquid flow must necessarily affect the final state of the reactor and although in general expansion occurs to about 5% of bed volume, some workers have operated reactors described as expanded beds at expansions of up to 25%.

8.1 Expanded Bed Reactors

8.1.1 Design and Operation

The anaerobic attached-film expanded bed (AAFEB) process was developed by Jewell and others as an extension of existing anaerobic digester systems in 1974. The AAFEB was considered to be the first anaerobic conversion system with the capacity to treat dilute wastestreams at ambient temperatures; Jewell [2] attributed this ability primarily to two main factors:

1) the formation of an active biomass concentration of around 30 kg m^{-3} of reactor, which was capable in theory of reaching concentrations of the order of 100 kg m^{-3}; and,
2) the trapping and filtering of fine inert particles which could result in a final effluent SS concentration of less than 5 mg l^{-1}.

These determinants combined to provide an efficient and high rate biological conversion process which did not require the high fluid flow rates necessary in comparable aerobic systems and which produced low sludge volumes for disposal.

The AAFEB reactor is typically a cylindrical structure packed with inert supportive particles to about 10% of its volume. The media may comprise sand, gravel, anthracite or plastic [3] to which the anaerobic bacteria composing the biofilm attach. The media particles, of slightly larger diameter than those employed in fluidised bed reactors (0.3–3.0 mm), are covered in the biofilm matrix and expanded by a vertical fluid velocity which is generated by a high degree of recycle. The expansion of the bed is sustained at a level at which each carrier particle retains its position relative to every other particle within the bed. A settled bed of particles under no external applied force will experience an initial expansion of up to 5% when a fluid flow is applied upward through the bed; a more loosely-packed arrangement results [4]. The phenomenon of channelling may cause limited localised displacement but overall the media grains remain in stationary contact. Further expansion of the bed results from biomass growth and the generation of biogas within the bed which accumulates in the interstitial spaces. If the upward flow to such an expanded bed were to be increased to the point at which the trans-bed fall in pressure was just equivalent to the weight of the media in the bed (corrected for buoyancy in liquid), then the carrier particles would be in free suspension in the upward fluid flow i.e. at the point of minimum fluidisation, and any additional increase in fluid flow velocity would result in the conversion of an expanded to a fluidised bed. In their investigations, Schraa and Jewell [5] utilised a bed expansion of 10–20%, adjusting the recycle pump rate to

122

compensate for the expansion caused by biomass variation. The major advantage in expanding the anaerobic reactor bed appears to be the minimisation of clogging problems whilst simultaneously accumulating significant quantities of bacteria on the surfaces of the supportive particles. Blockages however, are reported to occur to some extent [5].

The energy input necessary for expansion of the media in an AAFEB system is high, and in order to reduce it the density of the component carrier grains is required to be as low as possible but yet sufficient to permit rapid disentrainment from upflowing liquid. Biomass growth on the carrier does however increase overall particle density. Many of Jewell's more recent investigations have been performed with porous alumina, polyvinyl chloride plastic materials and ion exchange resin [7, 8]. The pores of the carrier alumina particles ranged from between 100 and 400 Å, and for this reason the microorganisms could not gain access to the internal structures of the support material: biological activity was ascribed to those bacteria comprising the surface biofilm and also to free cells in the fluid phase.

The AAFEB typically supports a strict biofilm structure and porous inert carrier particles with pores of dimensions sufficient to support entrapped microbial growth promote film development. It was found possible to maintain a mature, attached biofilm at 55 °C, with microbial attachment in excess of 90% for periods up to one year [7]. Biomass concentrations of 11 and 13 g VSS 1^{-1} were achieved in two separate reactor systems. The support material employed was diatomaceous earth, the porosity of which (when clean) was 67%. Within 2.5 months of the start-up of the AAFEB units a well-developed microbial film had grown up in each, of approximately 20 μm in thickness. Both methanogenic and non-methanogenic bacteria were present in the attached and entrapped biomass. At higher organic loading rates, increased biomass concentrations occurred. This phenomenon was probably attributable to the increased levels of substrate available to the bacteria; active biofilm volume has been observed to increase with increasing substrate concentration [9]. Maximum biomass concentrations of 60 g VSS 1^{-1} were recorded by Schraa and Jewell [5], of which 90% ($\pm 5\%$) of the total was attached. The thickness of the biofilm tended to increase with loading rate although evidence from electron microscope examinations of variable thicknesses of biofilm have shown that increase in film depth is a consequence of the excess production of extracellular polymeric substances (i.e. the matrix of the film) rather than an increase in actual microbial numbers [10]. The latter phenomenon may however be dependent upon factors other than organic loading, such as waste composition and the bacterial populations present in the anaerobic systems: slime-formation is enhanced by the addition of oligo- and polysaccharides but bacterial extracellular materials have been found to contain other components (see Chap. 3).

Almost all the reactions involved in a digestion system converting sucrose [5] appeared to be performed by the attached biomass, the fraction entrapped within the pores of the support media playing a minor role. In the treatment of a domestic wastewater the presence of a large population of anaerobic protozoa was reported [8]; they were believed to play a part in the bioconversion process, although their exact functions were not elucidated.

The biomass concentration of a fully-operational anaerobic expanded bed is typically greater than 20 kg VS m^{-3}, and it has been suggested that if the ultimate biomass concentration of the AAFEB were to approach 100 kg m^{-3}, the system could operate at removal rates of the order of 50 kg COD m^{-3} whilst maintaining SRTs in excess of 30 days [2]. Film thickness is likely to be governed to some extent by sloughing of attached film by bubble generation within it and by external shearing forces. Film and/or bacterial characteristics may be altered by loading fluctuations and biomass structure can be damaged by significant amounts of suspended organic matter in the influent wastestream. Fixed-biofilm processes are typically stable to a high degree when compared with conventional anaerobic digestion systems, a property attributed to the protection afforded against environmental influences by the surrounding matrix and the settling characteristics of the component carrier particles. Washout of particulate matter by rising gas bubbles and subsequent reduction of bioactivity and deterioration of effluent quality can nevertheless occur.

The recycle rates of full-scale AAFEB reactors are necessarily high and channelling in these systems tends to be a problem. Foaming at the top of the reactor can cause particle escape and mechanical or hydraulic controls may have to be implemented to minimise this phenomenon. The risk of active biomass loss during periods of process failure must also be considered and separation features such as ultrafiltration may be necessary for some reactor assemblies [2]. One reactor configuration reported included a tube settling device to control the effluent SS [8]. The apparatus had little effect on the parameters observed in the treatment of dilute wastewaters; a simplified reactor design was thus substituted in succeeding experiments and a recycle rate of the order of 100 ml min^{-1} maintained the bed expansion at the required level. The media volume increased during biomass growth and the clarifier region of the AAFEB, i.e. that reactor space above the effluent port, was cleaned daily of the small quantities of solids that were carried up by bubble formation. The majority of solids were reported to remain in the reaction zone and the additional portion of the vessel was deemed unnecessary for solids control. Effluent SS in AAFEB systems have been reported to increase only slightly for significant increases in hydraulic loading rate [7] and this problem is therefore not of major concern in the majority of expanded bed processes.

In an experiment designed to permit the definition of parameters which would cause system failure, HRTs as low as 5 min were applied to expanded bed processes [8]. At an HRT of 30 min the AAFEB consistently removed as much 80% COD from primary settled wastewater of average strength 186 mg COD l^{-1}; at 5 min HRT the effluent COD concentrations approached those of the influent, representing the point of system failure, but the biological microstructure was not destroyed as biomass and particle washout did not immediately ensue.

The mixing characteristics of anaerobic reactors appear to have an effect upon their operational performances: lithium tracer studies undertaken [11] indicated that the AAFEB exhibits a large degree of dispersion of substrate and mixing properties were defined as intermediate, whith a bias towards good. The effectiveness of the expanded bed process may be accredited to the large surface area to volume ratio made possible by the use of small carrier media, the relatively thin nature of the biofilm minimising diffusional difficulties and the large mass of attached bacteria that can be maintained within the bed at high fluid velocities.

8.1.2 Process Efficiency

The AAFEB reactor system has been employed for the treatment of a number of wastes including molasses fermentation [12] and domestic [2, 8, 13] and synthetic effluents [7]. Switzenbaum and Jewell [7] considered the expanded bed configuration to be effective for the conversion of low strength organic substrates (COD < 600 mg l^{-1}), operating at low temperatures of 10–20 °C, at high organic loading rates (> 8 kg COD m^{-3} d^{-1}) and at short HRTs in the region of several hours. These results were substantiated by other investigators [14] who further suggested that acclimation of the reactors to the influent wastestream was an additional factor in the enhancement of system stability under shock-loading and other adverse conditions. However, dilute domestic sewage, the TOC of which varied between 40 and 260 mg l^{-1} was applied to an anaerobic expanded bed, and the removal achieved found to be low, of the order of 17–52% [13] except at loadings below 0.5 kg TOC m^{-3} d^{-1}; the investigators therefore concluded that under typical real-scale conditions the AAFEB would not be suitable for wastestreams of such a nature.

The expanded bed configuration has been operated with a variety of wastes including dilute cow manure, glucose solutions [8], particulate cellulose [15] and whey [16] as well as domestic sewage and those feedstocks mentioned previously. Whey conversion by the AAFEB was reported to produce a maximum COD removal efficiency of 80%, utilising a bed expansion of 25% at HRTs of 27–4 h, organic volumetric loading rates of 8.9–60.0 kg COD m^{-3} d^{-1} and temperatures in the range of 25–31 °C [16]. Reactor pH was found to decrease at higher loading rates. This was coupled with lower organic removal efficiencies and indicated organic overload and reactor instability. The pH drop was probably the consequence of increased rates of VFA production by the acetogenic bacteria, which could not be matched by similar increases in the activity of acid-assimilating methanogenic bacteria.

AAFEB systems are also claimed to be capable of accommodating particulate organic matter in the form of pure cellulose [15]. Removal of 85% of the total influent COD was reported at loading rates of the order of 6.0 kg COD m^{-3} d^{-1}. Although particle size differed for each of the three reactors tested, the loading rates tolerated were observed to be similar, with hydrolysis of the influent solids being the rate limiting step in the overall reaction. It was also noted that hydrolysis occurred primarily within the entrapped rather than the attached biomass, enabling the solutes produced by the enzymic reactions to be assimilated by the bacteria of the film. This relationship may have been symbiotic with respect to the microbial populations: the position of the entrapped flora may have enabled more advantageous production of the extracellular enzymes required for hydrolysis while rapid assimilation of hydrolysis products by attached bacteria would permit the diffusion of fresh substrate to the hydrolysers.

The effect of temperature on the process efficiency of AAFEB reactors has been investigated. Using sugars as influent substrates, Schraa and Jewell [5] operated thermophilic expanded beds at 55 °C. Although start-up problems have been reported in AAFEB units, these investigators reported optimal growth conditions after 2.5 months of operation, attributing the superior process performance achieved to high biomass concentrations and the high reaction rates of thermophilic organisms. These types of reactor have been reported to withstand severe hydraulic overloading of raw wastewater [15] and the combination of high processing rates at increased temperatures and the accumulation of high biomass concentrations in the films is considered to be extremely promising for the conversion of warm and concentrated industrial wastes in small digesters with short HRTs, although reactor failure subsequent to exposure to toxic materials at elevated temperatures may occur [5]. Start-up of fixed-film reactors at temperatures in excess of 35 °C has been reported to be problematical due to the sensitivities of digester organisms, particularly the methanogenic bacteria, outside the mesophilic range [18] but Schraa and Jewell [5] obtained a total COD removal efficiency of 70% at a volumetric organic loading of 30 kg COD $m^{-3} d^{-1}$ with both medium (1.5–3.0 g COD l^{-1}) and high (5.0–16.0 g COD l^{-1}) strength soluble wastes. Switzenbaum and Jewell [7] found that decreasing temperature resulted generally in a decrease in process efficiency, but due to the extremely large biomass accumulation in fixed-film systems and the high SRT values realised, the AAFEB was demonstrated to compensate adequately for temperature fluctuations.

Variations in temperature, flow rate and influent concentration were reported to have little significant effect on the performance of one system of AAFEB reactors [14]; over a range of loadings from 0.3–24 kg COD $m^{-3} d^{-1}$ the COD removal rarely fell below 90%. However, in a domestic wastewater treatment investigation by Jewell [2], expanded beds at 20 °C were found to be less effective at increased COD loadings. COD loadings within the range of 0.65–35 kg $m^{-3} d^{-1}$ were applied to the system but above 4 kg $m^{-3} d^{-1}$ treatment efficiency rapidly deteriorated; less than 50% removal was achieved at 19 kg COD $m^{-3} d^{-1}$.

The effects of shock loadings of AAFEB units were examined in terms of reactor performance [8]. At a decrease in temperature from 35–10 °C, concomitant with an increase in loading rate from 1.3–24 kg COD $m^{-3} d^{-1}$ and an HRT decrease from 9.5–0.5 h, the COD removal efficiency decreased from 90% to around 45% and a significant proportion of the attached biofilm was lost. Despite these extreme conditions however, the total amount of biomass removed was reported to be a small fraction of the whole and process efficiency was restored in six days.

An AAFEB system treating domestic sewage containing between 50 and 600 mg COD l^{-1} required as little as 30 min HRT to achieve an effluent containing less than 40 mg COD l^{-1} [8]. The process was reported to be effective in the temperature range 10–35 °C. The operating parameters of the expanded bed could be drastically altered with little permanent adverse effect: in one instance, air was inadvertently pumped through the system for a 30 h period but upon adjustment to proper operating conditions, CH_4 production commenced almost immediately, indicating that the oxygen-sensitive methanogens had survived prolonged exposure within the protection of the biofilm.

126

Upon examination of the effects of temperature, influent substrate concentration, organic loading rate and HRT on the treatment capacity of the AAFEB, it was observed that substrate removal efficiency was primarily a function of both HRT and organic loading rate, although some degree of influence was also exerted by temperature and substrate concentration [7]. Over 80% COD removal was achieved at low HRTs and high organic loading rates regardless of influent substrate composition or temperature. Hence, the AAFEB is an anaerobic treatment process capable of accommodating a number of moderate and high strength wastestreams at both mesophilic and thermophilic temperatures, at short HRTs and with great stability under shock-loading conditions. Most of the advantages of the system can be attributed to the high concentration of biomass achieved on the small carrier media employed and the rates of recycle which render the digestion process less liable to upset by toxic components of the influent.

8.2 Fluidised Bed Reactors

8.2.1 Design and Operation

The operational characteristics of anaerobic fluidised bed (AFB) reactors have been outlined by Cooper and Wheeldon [4] and Heijnen [1]. The biological fluidised bed concept was initially applied to denitrification processes and subsequently to carbonaceous and nitrogenous oxidation systems [19, 20]. The utilisation of anaerobic fluidised beds for denitrification of both low and high nitrate concentrations in sewage and nuclear fuel processing wastes has been reported by Francis and others [21, 22].

The influent feed to the typical AFB enters at the base of the reactor (see Fig. 24). Even flow distribution through the reactor cross-section, especially in the zone close to the feed entry point, can, however, present a design problem in fluidised bed systems (as it also does in packed bed units) and some form of feed distributor is requisite. In one investigation, a layer of gravel at the reactor base was employed [23] but this proved unsatisfactory due to frequent blockage. It was therefore replaced by a conical distributor; the influent was orientated downwards into the base of an angled cone, resulting in a dynamic upwash in the lower part of the vessel. The circulatory flow pattern of an AFB unit has been observed to be smoothed out by the incorporation of a perforated distribution plate above the cone [4]; the height of the influent upwash was limited by a stainless steel distributor plate with 6 mm diameter holes, the conical section of the bed functioning in effect as the distributor for the fluidised bed which began where the recirculatory pattern subsided. The carrier particles of the conical section exhibited little biomass accumulation and a distinct interface above the distributor was formed, where typical biomass growth occurred. The high flow rates employed in AFB reactors promote optimal contact between biomass and wastestream and are necessary for effective COD removal, especially where dilute wastestreams are treated, as the low substrate concentrations present tend to result in reduced rates of COD removal [24].

The bacteria in fluidised bed systems attach to small diameter media which may be sand [23, 25], activated carbon, garnet or glass beads. The high vertical velocity of the wastestream to be treated expands the bed to a point beyond which the net downward gravitational force is equalled by the frictional drag. An extremely high degree of recycle is required and single carrier grains do not have a fixed position within the bed (cf. the expanded configuration) but each particle tends to remain nonetheless within a restricted location. Other anaerobic digesters such as the packed bed (anaerobic filter) assembly must contain large supportive grains or microbial growth quickly clogs them: the surface area of bacterial film per unit vessel volume of these reactors, and consequently the volumetric productivity, are thus low. In an AFB, such mechanical constraints do not apply as the carrier particles can be much smaller: the reaction zone expands to accommodate increasing bacterial growth. Support particles typically tend in addition to be spherical or quasi-spherical, as these forms have advantages associated with ease of fluidisation and the understanding of their kinetic characteristics under fluid-flow conditions is extensive.

Inert support particles of sizes 0.5 mm and 0.7 mm diameter commonly in use in AFB systems were compared as part of an investigation [26]. Equivalent performances of the reactors employed were attained in a shorter time from start-up using the sand particles of smaller dimensions; in addition, the unit operating with the 0.7 mm particles required twice the recycle rate of the reactor utilising the smaller material for a bed expansion of 30–50%. Switzenbaum et al. [25] substantiated these observations by demonstrating that reactor efficiency was improved by the replacement of 0.5 mm diameter sand with particles of 0.2 mm diameter. The large surface area that can be achieved by the use of small media and the degree of fluidisation which permits the entire surface area of each grain to become available for bacterial colonisation results, therefore, in surface areas of 300 m^2 m^{-3} in a typical fluidised bed reactor. Biomass concentrations markedly superior to those found in conventional systems, of the order 8–40 g VSS l^{-1} can be achieved; reactor size and treatment times are hence significantly reduced [27].

The high biomass volume possible in fluidised bed systems must be controlled for maximal performance of the units. The growth of microorganisms in balls of wire has been claimed to give finer control over the size and density of fluidised bed particles [28] and a biofilm attached to activated carbon particles has been shown to utilise substrate previously adsorbed onto the carbon [29]. This has suggested the possibility of supplying the film with nutrient both from the surface and base at start-up, thus minimising any initial diffusional limitations, although the thin nature of the matured biolayer probably eliminates this difficulty. The biofilm structure is itself distinct and its form is considered to be influenced to a marked extent by the composition of the carrier media [30]. Bacterial activity in fluidised beds, unlike that in other fixed-film reactors such as anaerobic filters, has been shown to be greatest for both acetogenic and methanogenic bacteria in the central region of the bed [31]. The mobility of the supporting medium and the high degree of recycle required for fluidisation were believed to influence this phenomenon. In the packed bed reactor for example, microbial activity and substrate profiles have indicated that the majority of waste conversion occurs in the lower region of the filter [32–34].

Several factors have been claimed to contribute to the superiority of the AFB system, all of which may be attributed to a greater or lesser extent to high biomass concentration:

1) maximum contact between the biocoated media and the liquid phase comprising the wastestream to be treated occurs;
2) liquid-film diffusional resistances due to particle motion and liquid velocities are minimal;
3) channelling, plugging and gas hold-up problems are generally circumvented;
4) bacterial film thickness can be controlled and optimised; and,
5) high biomass concentrations reduce the reactor size and hence land area necessary for the treatment plant.

The latter consideration renders the use of the AFB as an on-site treatment process highly effective; the land area requirement can be as much as 80% less than that typical of conventional anaerobic units [6].

The problem of cell washout that complicates the operational stability of conventional stirred tank and other fermenter assemblies is less likely in an AFB reactor. According to Andrews [35] washout is impossible as long as the superficial liquid velocity is kept below the settling velocity, although this has proved to be extremely difficult in practice. The fixed-film nature of the fluidised bed process permits the maintenance of extended mean cell residence times at low HRTs without the requirement for biomass settling and recycle; organic loading and bacterial growth rate are the only two parameters controlling the mean cell residence time. A biomass concentration of 5.65 kg VS m^{-3} has been reported for an AFB reactor with an SRT of 12.0 days and an SRT/HRT ratio of 23 [36].

A wide range of operating conditions, together with improved reactor stability was claimed for a fluidised bed design that made use of a tapered reaction vessel [37]. The unit resembled an inverted truncated cone which gradually expanded from the relatively small cross-sectional area of the influent entry point to an area several times greater. The flow patterns generated in the reactor were reported to exhibit minimal backmixing, especially at the feed entry point and few large eddy currents were observed, fluid velocity within the bed decreasing with reactor height. At increased flow rates the lower portion of the biomass plus the supportive media simply expanded into a portion of the vessel having a greater cross-sectional area. Under high fluid flow conditions the lower zone of this type of system may be relatively free of the fluidised bed itself, as the liquid velocity may greatly exceed the settling rate of the coated particles comprising the bed at that point. The effects of such a design and mode of operation upon microorganisms existing under these conditions have not been elucidated. Some phase separation of bacterial populations must inevitably be the outcome of variable conditions in different regions of the bed. Bull et al. [31] recorded partial phase separation in an AFB at a COD loading of 12 kg m^{-3} d^{-1}, methanogenic activity peaking at a slightly higher point in the bed than acetogenic activity; this was confirmed by a VFA profile through the system, although at lower organic loading rates phase separation was not observed. The effect may have been purely physical, gas production by the methanogens within the biofilm causing the selective rise of coated particles as gas bubble disentrainment occurred.

A relationship has been proposed [38] which is capable of predicting the height of an AFB in relation to the extent of the biomass growth at the surface of the fluidised particles. The application of this relationship in the determination of one of the parameters of wastewater treatment (the organic carbon density of bacterial film) was demonstrated. A number of investigators have proposed relationships applicable under certain restrictive conditions between the height of the fluidised bed within the reaction vessel and the coverage of biofilm on the support particles [39–41]. A new mathematical model for use in the design and optimisation of AFB systems in which the biomass particle size was not a required input parameter but was predicted as a consequence of the process by which the reactor reached steady state has also been presented [35]. The simplified mathematical model of Mulcahy et al. [42] identified system variables and was suggested as a basis for process optimisation. The first order substrate assimilation model of Chen and Hashimoto [43] was reported to provide an accurate prediction of the reactor performance of a system treating high strength whey permeate [44]. Rittmann and McCarty [45, 46] investigated biofilm kinetics in detail and used their analyses to compare four reactor configurations [47]. The model predicted the superiority of a single-pass fluidised bed, as the plug-flow liquid regime aided the even distribution of biofilm throughout the reactor. The addition of recycle to an AFB was considered to reduce performance efficiency to levels approaching that of completely mixed units. The recycle mode of operation in fluidised beds does however provide a great advantage in the treatment of potentially inhibitory wastes; the high fluid rates employed ensure short exposure to damaging materials whilst the recycling mode allows efficient breakdown of substrate.

8.2.2 Process Efficiency

Wastewaters which have been treated by AFB systems include meat and dairy wastes, whey permeate, food processing, bakery, chemical, soft drink bottling and acid whey wastes, beet molasses and heat treatment liquor [23, 27, 36, 48].

A pilot-scale AFB reactor was employed for the conversion of whole whey effluent from a cheese manufacturing plant; the system was operated at 35 °C [27]. Loading rates were applied of between 13.4 and 37.6 kg COD $m^{-3} d^{-1}$ and subsequent COD reductions of 83.6% and 72.0% respectively were achieved, which represented HRTs of between 1.4 and 4.9 days. At 24 °C COD removals were reported to be of the order of 10% less than those obtained at the higher temperature, being 71.0% and 65.2% at organic loading rates of 15.0 and 36.8 kg COD $m^{-3} d^{-1}$ respectively. High strength organic wastewaters (synthetic meat and dairy waste) were effectively treated in an AFB reactor at COD loading rates of up to 6 kg $m^{-3} d^{-1}$ at 37 °C and 3 kg $m^{-3} d^{-1}$ at 20 °C [23]. Effluent SS proved to be greater in comparison to those produced by reactors treating dilute wastes and using lighter support particles, a factor attributable to the higher upflow velocities necessary for fluidisation. However, COD reductions in excess of 70% could be achieved.

Moderately high strength chemical waste comprising mainly ethanol was also treated using a fluidised bed [27], with organic loading rates in the range 4.1–

27.3 kg COD m^{-3} d^{-1} applied. COD removal efficiencies of between 93% and 79% with 81–84% CH$_4$ content in the biogas were achieved. Food processing effluent and heat treatment liquor were effectively dealt with in a system comprising pilot-scale AFB reactors, although soft drink bottling waste, high in refined sugars, was degraded more slowly. This latter effect may have been caused by bacterial inhibitors present in the waste itself: addition of such product preservatives is not uncommon in food and drink manufacturing processes.

Soluble COD removal rates of up to 19 kg m^{-3} d^{-1} were reported in a fluidised reactor treating high strength lactic casein whey permeate (2–7 g l^{-1} soluble COD) at 35 °C [44]. A lower removal efficiency of 50% at an organic loading rate of 3.0 kg soluble COD m^{-3} d^{-1} was manifested at 15 °C but the AFB system proved to have a much lower nutrient requirement than that necessary in an equivalent CSTR unit, and no supplemental nitrogen or phosphorus was added.

Variations in loading and temperature are reported to be readily accommodated by AFB systems. A strongly organic whey waste applied to a reactor at loadings of up to 37.6 kg COD m^{-3} d^{-1} resulted in increased VFA concentration (and hence depressed pH), but no inhibition of the reactor performance was recorded and the average CH$_4$ content of the biogas was maintained at 60% [27]. The AFB volatile acids concentration was found to be highly dependent upon organic loading rate, a situation probably arising from the variable rates of substrate assimilation of the various bacterial populations within the reactor. An increase in organic loading resulted in an increase in soluble COD removal rates in the system of Boening and Larsen [44] but overall percentage removal efficiency decreased. Biomass concentration was observed to increase with decreasing temperatures (35–12.5 °C), compensating for the reduced reaction rates of bacteria at lower temperatures and thus reducing the sensitivity of the system to temperature change.

The stability of an unheated (20 °C) AFB reactor was investigated: the unit was operated on synthetic meat waste as feedstock and the effects of COD, pH, hydraulic and temperature shocks were examined [49]. The reactors responded satisfactorily to transient increases in influent flowrate: at an initial COD loading of 2.6 kg m^{-3} d^{-1}, flowrate increases of 100% and 150% for 4 and 8 h periods resulted in an immediate decrease in reactor pH of 0.15 and 0.25 units, concomitant with an increase in VFAs and an increase in effluent COD values of 40–50%. Raising the influent COD by 150% for 4 and 8 h periods caused a similar pH drop, with increased effluent COD, SS and VFA. A 300% increase in influent COD for 4 h, however, produced raised levels of alkalinity, COD and VFAs for up to 3 h after return to normal reactor conditions. Increased levels of effluent VFAs upon increase of influent COD have also been reported by Stephenson and Lester [36]; these were considered to be the result of accumulation of higher fatty acids such as propionate and butyrate, which are frequently present under overload conditions [3].

The only parameter significantly affected by a reduction in pH to 3.0 for a period of 8 h in an AFB was the alkalinity, which decreased [49]. The effluent pH dropped by only 0.1 unit. Shock loading of an AFB with orthophosphoric acid-supplemented feedstock, which reduced influent pH to 3.0, for a period of 36 h, caused a significant decrease in effluent pH although effluent SS and VFAs remained steady [36], indicating the stability of the system to toxic shock; a CSTR system under similar adverse conditions ceased to function after 24 h.

The treatment of low strength wastewater by AFB systems, however, may require high pumping rates to achieve effective COD removals; these may not be economically offset by the advantages of the fluidised system. Matsui et al. [50] reported that COD removals were low (of the order of 23.8%) when a dilute waste was applied to an AFB, but the COD loadings examined were very high, being in the range of $18.6–131.6 \text{ kg m}^{-3} \text{ d}^{-1}$. Primary effluent municipal wastewater has been pre-treated effectively on a pilot-scale AFB reactor under unsteady state conditions, and the results appear encouraging [25].

The advantages of anaerobic treatment of moderate and high strength wastes, i.e. production of utilisable CH_4, low cell synthesis, no oxygen requirement and low nutrient requirement, are well known. However, the technical limitations of conventional digester processes – large reactor volumes and HRTs, intolerance of rapid temperature changes and high organic loading rates – are substantially reduced or eliminated by the use of a fluidised bed configuration. The extremely high biomass concentrations that may be achieved through the use of small-diameter carrier media, in conjunction with the high rates of recycle typically employed in AFBs, enable this type of reactor to accommodate severe overload conditions, alterations in influent composition and temperature changes with great efficiency. The potential of AFB systems for the conversion of high strength soluble industrial wastewaters has been demonstrated, with low HRTs and thus high processing rates and high quality effluent being among the major advantages, in addition to reduced reactor volumes and consequently much smaller land area requirements than those necessary for conventional waste treatment processes.

References

1. Heijnen JJ (1983) Development of a high-rate fluidised bed biogas reactor. In: Proc Eur Symp Nov, Noorwijkerhaout, Netherlands, p 259
2. Jewell WJ (1981) Development of the attached microbial film expanded bed process for aerobic and anaerobic waste treatment. In: Cooper PF, Atkinson B (eds) Biological fluidised bed treatment of water and wastewater. Ellis Horwood, Chichester, p 251
3. Henze M, Harremoes P (1983) Water Sci Technol 15:1
4. Cooper PF, Wheeldon DHV (1981) Complete treatment of sewage in a two-fluidised bed system. In: Cooper PF, Atkinson B (eds) Biological fluidised bed treatment of water and wastewater. Ellis Horwood, Chichester, p 121
5. Schraa G, Jewell WJ (1984) J Water Pollut Control Fed 56:226
6. Cooper PF, Wheeldon DHV (1980) Water Pollut Control 79:286
7. Switzenbaum MS, Jewell WJ (1980) J Water Pollut Control Fed 52:1953
8. Jewell WJ, Switzenbaum MS, Morris JW (1981) J Water Pollut Control Fed 53:482
9. Trulear MG, Characklis WG (1982) J Water Pollut Control Fed 54:1288
10. Salkinoja-Salonen MS, Nuys E-J, Sutton PM, van den Berg L, Wheatley AD (1983) Water Sci Technol 15:305
11. Forster CF, Rockey JS, Wase DAJ, Godwin SJ (1982) Biotechnol Lett 4:799
12. Frostell B (1980) Wastewater treatment and energy recovery in an expanded bed system. In: 3rd Int Congress on Ind Wastewater and Wastes, Stockholm
13. Rockey JS, Forster CF (1982) Environ Technol Lett 3:487

14. Jewell WJ, Morris JW (1981) Influence of varying temperature, flowrate and substrate concentration on the anaerobic attached-film expanded-bed process. In: Proc 36th Ind Waste Conf, Purdue Univ, Lafayette, Indiana 1981. Ann Arbor Science, Ann Arbor Michigan, p 655
15. Morris JW, Jewell WJ (1981) Organic particulate removal with the anaerobic attached-film expanded-bed process. In: Proc 36th Ind Waste Conf, Purdue Univ, Lafayette, Indiana 1981. Ann Arbor Science, Ann Arbor Michigan, p 621
16. Switzenbaum MS, Danskin SC (1981) Anaerobic expanded bed treatment of whey. In: Proc 36th Ind Waste Conf, Purdue Univ, Lafayette, Indiana 1981. Ann Arbor Science, Ann Arbor Michigan, p 414
17. Kennedy KJ, van den Berg L (1982) Water Res 16:1391
18. Salkinoja-Salonen MS, Hakulinen R, Valo R, Apajalahti J (1983) Water Sci Technol 15:309
19. Jeris JS, Beer C, Mueller JA (1974) J Water Pollut Control Fed 46:2118
20. Jeris JS, Owens PW, Hickey R, Flood F (1977) J Water Pollut Control Fed 49:816
21. Francis CW, Callahan MW (1975) J Environ Qual 4:153
22. Francis CW, Malone CD (1977) Progr Water Technol 8:687
23. Bull MA, Sterritt RM, Lester JN (1982) Trans I Chem E 60:373
24. Pohland EG, Verstraete W, Cohen A, Goma G (1981) Process design. In: 2nd Int Symp on Anaerobic Digestion, Sept 6–11, Travemünde, Germany
25. Switzenbaum MS, Sheehan KC, Hickey RF (1984) Environ Technol Lett 5:189
26. Sutton PM, Li A (1983) Water Sci Technol 15:333
27. Hickey RF, Owens RW (1982) Biotechnol Bioeng Symp 11:399
28. Atkinson B, Black GM, Lewis PJS (1979) Biotechnol Bioeng 21:193
29. Andrews GF, Tien C (1981) AIChE J 27:396
30. Murray WD, van den Berg L (1981) Appl Environ Microbiol 42:502
31. Bull MA, Sterritt RM, Lester JN (1984) Water Res 18:1017
32. Young JC, McCarty PL (1969) J Water Pollut Control Fed 41:RI60
33. van den Berg L, Lentz CP (1979) Comparison between up and downflow anaerobic fixed film reactors of varying surface-to-volume ratios for the treatment of bean blanching wastes. In: Proc 34th Ind Waste Conf, Purdue Univ, Lafayette, Indiana 1979. Ann Arbor Science, Ann Abor Michigan
34. Young JC, Dahab MF (1983) Water Sci Technol 15:369
35. Andrews GF (1982) Biotechnol Bioeng 24:2013
36. Stephenson T, Lester JN (1986) Biotechnol Bioeng (in press)
37. Scott CD, Hancher CW (1976) Biotechnol Bioeng 18:1393
38. Wang S-C, Tien C (1983) Can J Chem Eng 61:64
39. Andrews GF, Tien C (1977) AIChE J 25:270
40. Tsezos M, Benedek A (1980) Water Res 14:689
41. Shieh WK, Sutton RM, Kos P (1981) J Water Pollut Control Fed 53:1574
42. Mulcahy LT, Shieh WK, Lamotta EJ (1981) AIChE Symp 209(77):273
43. Chen M, Hashimoto K (1979) Biotechnol Bioeng Symp 8:269
44. Boening PH, Larsen VF (1982) Biotechnol Bioeng 24:2539
45. Rittmann BE, McCarty PL (1980) Biotechnol Bioeng 22:2343
46. Rittmann BE, McCarty PL (1980) Biotechnol Bioeng 22:2359
47. Rittmann BE (1982) Biotechnol Bioeng 24:1341
48. Frostell B (1982) Process Biochem 17:37
49. Bull MA, Sterritt RM, Lester JN (1983) J Chem Technol Biotechnol 33B:221
50. Matsui S, Shibata T, Imai A (1979) Water Purific Liquid Wastes Treat 20:1117

9 Developments in Reactor Design

Anaerobic systems tend to be subject to a number of operational limitations. More sophisticated reactor designs such as expanded and fluidised beds and the sludge blanket configuration often require careful monitoring and management. A number of difficulties arise from deficiencies in the knowledge of the microbiological and biochemical reactions of anaerobic systems; these include biofilm build-up, pH and micronutrient levels, temperature effects, sensitivity to toxic compounds and acclimation characteristics of digester organisms. Additional problems occur due to the lack of knowledge of many aspects of the digester system as a whole, primarily with reference to the start-up and reliability of the various reactors. Despite the research in progress, practical experience of pilot or full-scale systems has been limited. Anaerobic digester operations vary widely in complexity and configuration and no invididual design can be regarded as ideal, as each application requires a particular system to fulfil specific criteria. Efficient on-line monitoring of process parameters must likewise be developed for efficient operation.

9.1 Improvements

The control parameters used in the latter stages of anaerobic digestion to monitor the state of the system are typically the rate and composition of the biogas evolved and reactor pH. However, instrumentation has recently been developed to register on-line the redox potential of the liquor and the partial pressure of the H_2 component of the gas phase [1]. The latter parameter seems particularly promising and effective control models based on this may become available in the future. The development of a simple computer control system to alleviate the need for analytical equipment has been refined for use with anaerobic downflow stationary fixed-film reactors [2]. Gas production was chosen as the controlled variable and feed rate as the manipulated variable. The control system managed a reactor for several months. The use of a microcomputer in the measuring and control of laboratory fermenters has also been reported [3] and a simple, inexpensive piece of apparatus for the continuous pH measurement of anaerobic mixed liquors has been described [4].

In anaerobic reactor systems, especially those dependent upon unattached biomass, maximisation of sludge and/or floc activity and concentration provide means by which efficiency may be improved. There are several factors that affect sludge activity, the most obvious of which is the influence of the wastestream. An influent high in SS, particularly those of inorganic origin, will cause accumulation of sediment in the reactor and the consequent loss of digester volume can be substantial. Biomass activity may be further limited by the degree of diffusion of substrate into individual suspended flocs. The diffusion depth, based on kinetic parameters, has been calculated to be about 1 mm [5]. This indicates that sludge flocs, granules and also attached biofilms should not be greater than approximately 1 mm in thickness, although the sludge settling rate tends to impose a minimum limit on particle or floc size and problems arise where large granules are required.

The concentration of biomass in a waste treatment reactor must also be maintained for improved digestion performance and recirculation techniques are widely applied where washout tends to occur. In typical anaerobic degradation processes, the solids in the digester effluent are returned to the reactor vessel via an internal (e.g. sludge blanket) or external (e.g. contact process) settling unit. This modification allows extended SRTs to be maintained even under conditions of large hydraulic throughput (short HRT) and as a consequence, reactor size and associated cost have been reduced. The large, rapidly developing biological populations which are obtained, when combined with improved operational instrumentation, have increased process efficiency and reactor stability to system fluctuations.

The beneficial effect of sludge recycle on the concentration of sludge achieved is almost totally dependent upon the settling rate of the biomass. In the contact process, for example, the sludge is present as discrete flocs with SVI values of 70–150 ml g^{-1}; the settling properties of the sludge are functions of the wastewater components and the sludge load [5]. A high sludge load tends to produce bulking sludge, which decreases reactor efficiency. A certain degree of recycle produces significant benefits to the reactor system [6]. These include:

1) some neutralisation of the pH of the incoming wastestream;
2) reduction of the alkalinity required;
3) amelioration of the effect of biodegradable toxic compounds of the wastestream;
4) reduction of the effects of shock loadings; and,
5) compensation for fluctuations in influent flow rate.

In fixed-film reactor configurations, the choice of particle diameter and density affects the settling rate of the biofilm-covered particles. This rate can be exceedingly high e.g., 50 m h^{-1} [5] and can allow the maintenance of a concentrated suspension of film-covered media with a biomass accumulation of 30–40 g VSS l^{-1}, in addition to a high superficial liquid velocity of 10–30 m h^{-1} in the reactor. A number of supplementary factors pertaining to inert support particles in fixed-biofilm reactors are in the process of investigation, for further optimisation of these systems. A decrease in the media particle diameter of the expanded and fluidised bed bioreactors will increase the surface area and reduce the pumping cost for expansion or fluidisation of the medium and recycle. It has already been established that use of smaller-diameter particles (sand) decreases start-up time and increases efficiency [7]. Other factors influencing the choice of support media include density of media, density of biomass, waste specificity and cost; some of these may be altered and an optimal combination achieved.

The surface of carrier particles is generally two-dimensional, but some systems employ the three-dimensional configuration of porous support materials [8]. The bacterial cells immobilise on the internal walls of the carrier particles; up to 10^9 cells per g of ceramic carrier could be achieved if the pore diameters of the media were in the range of one to five times the length of the cell. Another important function of porous media was found to be the improved retention of biological solids; again, pore size was the dominant factor [9] as poorer performances resulted from reduced pore sizes. The efficiency of fixed-film systems is also dependent upon the degree of porosity of the carrier employed. A carrier with a pore volume of 75% was reported to be superior to those of 67% and 36% in one investigation [10], although the results of Wilkie and Colleran [11] contradicted this, with plastic media of high porosity (94%) giving a performance inferior to that of clay media of a lower (69%) porosity value. In the latter case, media composition was the major determinant of treatment efficiency.

The overall loading rates obtained in anaerobic digestion tend to be low compared with industrial fermentation processes such as ethanol production [12]. Under optimal conditions, methanogenic digestion of precursor substrates such as acetate can in practice reach CH_4 production rates of 20–30 $m^3\ m^{-3}\ d^{-1}$, corresponding to COD removal rates of 40–60 kg $m^{-3}\ d^{-1}$. In the ethanol fermentation process, rates expressed as kg of ethanol COD produced per m^3 of reactor per day can reach values of 2 000–3 000, i.e., fifty times greater. The methanogenic stage of anaerobic digestion in particular requires protracted substrate retention times, which in turn need larger digester volumes when conventional reactors are employed and in situations where, for example, sewage sludge and slurry loadings are only 1–2 kg COD $m^{-3}\ d^{-1}$.

In order to maximise the productivity of existing sludge digesters, the ratio of COD-particulate to soluble-COD can be altered, although increased particulate solubilisation is biologically as well as technically difficult [12]. The ratio of particulate to soluble organics can, however, be altered by the simple expedient of adding a second substrate to the primary wastestream. This procedure has been followed to alter the $C:N$ ratios of influent digester feedstocks [13, 14]; the enhancement of digestion by the addition of vegetable waste to animal excreta has already been mentioned (see Chap. 4). The use of cellulose as a co-substrate in the digestion of piggery manure allowed an increase of volumetric loading on a temporary basis in another investigation, the conversion to biogas being almost complete [15]. The addition of easily metabolisable concentrated wastes can also improve reactor efficiency. In a pilot scale examination into auxiliary substrates, a concentrated petrochemical wastestream (450 kg COD m^{-3}) mostly in the form of acetate, was added to sewage sludge at a 1 : 9 ratio [12]. The volumetric loading rate was increased two fold, and the corresponding rate of CH_4 production rose by a factor of four, with HRT remaining constant. COD reduction in a completely mixed reactor increased from 28–45% overall. Using the anaerobic contact process, a volumetric loading of up to 7.75 kg COD $m^{-3}\ d^{-1}$ was achieved with a COD reduction efficiency of 50%. In both reactors, approximately 90%

of the petrochemical COD was removed as biogas. According to Verstraete [12] the advantages of the above substrate-supplementation regime are that the anaerobic treatment plant facilities can be utilised more advantageously, nutritionally imbalanced wastes can be beneficially adjusted to provide improved growth requirements, and the more toxic components of one wastestream may be diluted to an acceptable level by the other.

The improvement of a conventional stirred tank digester by the addition of an inert material – in effect by converting it to a pseudo-fixed-film process – has been reported. Enhancement of CH_4 production by 10–15% was one of the improvements resulting from the addition of powdered activated carbon (PAC) to anaerobically digesting sludge [16]. The increased conversion of VS was also observed. It was suggested that the surface of the PAC provided adsorption sites for substrate accumulation and that the PAC could sequester toxic substances. Solids residence time could be increased by the preferential retention of sorbed VS and active biomass, resulting in a greater substrate conversion without alteration of HRT. The addition of PAC to anaerobic digesters is not a new phenomenon; a laboratory investigation and field studies were reported in 1935 by Rudolfs and Trubnick [17], indicating that poorly digesting sludge was stabilised, gas production was increased and a greater reduction in VS occurred when activated carbon was added to the system. Methane production in an anaerobic reactor dosed with carbon was found to be five times that of a control system after 3 weeks [18, 19]. Under stressed reactor conditions, an increase of 150% in CH_4 yield was observed upon carbon addition, the degree of increase appearing proportional to the carbon concentration added [20]. Improved operation of overloaded digestion systems following PAC addition was observed regardless of whether the stress resulted from toxic or hydraulic overload [21]. It has been indicated that PAC can enhance the dispersion of hydrophobic materials into a slurry and that such effects are increased at lower HRTs, suggesting that it may be used for increasing the capacity of existing anaerobic digesters [16]. Spencer [20] noted that carbon increased digester efficiency under certain operating conditions and listed several mechanisms to account for the enhanced digestion of substrate:

1) carbon provides the sites of anaerobic reactions;
2) carbon adsorbs inhibitory substances toxic to the digestion process;
3) the buffering capacity of the system is increased by the presence of alkaline carbons; and,
4) carbon supplies metal micronutrients necessary for the metabolism of anaerobic bacteria.

The analysis of test data in terms of mathematical models has indicated that the benefits of activated carbon addition may have resulted from localised changes in bacterial-substrate-enzyme interactions [16].

Although the above amendments to existing design processes may enhance overall efficiency of anaerobic digestion, the problem remains that a given set of reactor conditions cannot be optimal for the various types of digester bacteria, the degradation process being a multi-phase operation.

9.2 Multi-Stage Operations

The anaerobic digester in a single-stage operation acts as a continuous culture, with several microbiological conversion reactions occurring simultaneously. The residual growth limiting soluble substrate concentrations for these bacterial processes depend basically upon the K_{max} (maximum substrate utilisation rate coefficient) for the reaction and the detention time of the system, i.e., the growth rate of the bacterial cells. The levels of influent govern the concentrations of substrate in excess. The K_{max}, and thus the residual substrates, are low in sugar fermentations, for example, as single-stage anaerobic reactors are rarely operated at detention times of short enough duration to be close to the maximum growth-rates of the fermentative bacteria. At low detention times however, the residual acid concentrations in the system can be high, and these comprise a major proportion of the solubilised inhibitory compounds in digested sludges [14].

A continuous bacterial culture, having a second stage operating in series at a different detention time, although performing the same conversions as the initial stage, can theoretically catabolise primary stage residual substrates (see Fig. 25). In a simple sugar fermentation, it has been postulated that a two-stage process can supply greater substrate utilisation with a lower overall detention time than a one-stage reactor, as the secondary stage may be operated at a rate in excess

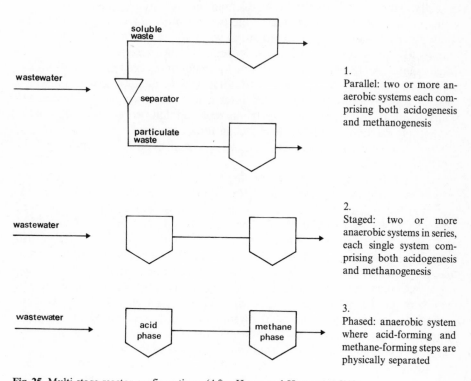

1.
Parallel: two or more anaerobic systems each comprising both acidogenesis and methanogenesis

2.
Staged: two or more anaerobic systems in series, each single system comprising both acidogenesis and methanogenesis

3.
Phased: anaerobic system where acid-forming and methane-forming steps are physically separated

Fig. 25. Multi-stage reactor configurations. (After Henze and Harremoes [22])

of the maximum bacterial growth rate without washout [14]. However, because of the complications of the sequential conversion steps in the digestion of slowly-degradable solid materials such as fibres, the primary stage must necessarily have an extended detention time and the resultant residual acids will be low regardless of the reactor conditions. The inevitable conclusion is that, for most digester systems dealing with such wastes, a second anaerobic stage would hardly be economical.

Anaerobic digestion followed by aerobic treatment has, however, been found to be economically advantageous and is well-documented for a number of wastes including coke-plant wastewater [23, 24] and sewage sludges [25–27]. Coal gasification wastewater was treated efficiently by the use of a three-stage series of reactors, the first two units of which were anaerobic [28]. Both anaerobic reactors were of the filter design, the first a jacketed polypropylene-packed column and the second a jacketed granular activated carbon (GAC) packed filter; both operated at 35 °C. The first reactor was removed midway through the investigation as little dissolved organic carbon (DOC) and COD reduction resulted. Appreciable DOC and COD removals throughout the duration of the test occurred in the GAC reactor, however, as it provided a larger sheltered microbial attachment area than the polypropylene configuration of the other anaerobic column, but virtually no reduction of cyanide, thiocyanate and ammonia were recorded. In contrast, another system comprising a primary anoxic and a secondary oxygenic phase was able to remove these generally inhibitory substances (see Chap. 5) successfully, with an HRT significantly less than that required by a completely oxygenic process [29]. Completely anaerobic two- or more stage processes have recently received attention. An investigation was undertaken to compare the operational efficiencies of three anaerobic systems [30]. An anaerobic fluidised activated carbon (AFAC) filter with cyclic replacement of the carbon media was compared with two two-staged systems, the first of which comprised an anaerobic filter of Raschig rings and an AFAC filter reactor in series and the second assembly three AFAC filters in series. It was reported that process inhibition in the single-stage reactor could be reversed by the addition of fresh carbon, but effluent of improved quality was obtained in the series systems, due to absorption polishing in the secondary stage.

A similar two-stage system comprising an anaerobic filter and a fluidised bed reactor in series, the filter containing Raschig rings and the bed GAC, was employed in the treatment of dilute coal gasification wastewater [29]. At loadings of 2.5 kg COD m^{-3} d^{-1} and a 2 day retention time, removals of 92% COD, 78% TOC, 93% phenol and 99% cresol were reported, with 3.6–8.1 l d^{-1} CH$_4$ evolution. The particular advantages of the system included a pilot scale configuration that permitted scale-up with a minimum of modification and the use of a first stage, but interchangeable, Raschig ring packed filter for initial acclimation and seeding of the second stage AFAC column. The fluidisation of the GAC column minimized gas retention, provided dilution and furnished buffering capacity. Glucose was added initially to the first reactor at the acclimation stage and the wastewater was supplemented with sodium and potassium phosphates, but vitamin and trace element additions were not considered necessary, although in a similar investigation, micronutrients were supplied to enhance acclimation [31].

Two anaerobic fluidised bed reactors in series, treating whey wastes, were reported to be much more efficient than a single reactor, with 94% COD removal efficiency at a loading rate of 10 kg m^{-3} d^{-1} and a 5 day HRT [6]. Whey wastes may have COD values in excess of 50 g l^{-1}; a single-stage fluidised bed system capable of 80% COD removal will produce an effluent which still contains 10 g COD l^{-1}. The series arrangement of Jeris [6] proved advantageous in that the 94% removal efficiency of the system reduced an influent COD of 52 g l^{-1} to 3 g l^{-1}. The advantages of a two-staged over a single-staged fluidised bed operation were summarised by Heijnen [5] as:

1) ease of start-up and operation;
2) very high treatment capacity; and,
3) no process instability due to sludge loss.

A dual-staged anaerobic unstirred tank digester system was utilised for the treatment of alcohol stillage derived from sugar cane molasses [32]. Gas production was 0.4 m^3 kg^{-1} COD converted, with 4 days HRT, as compared to the holding times of between 10 and 15 days with a single reactor, reported in the literature. The treatment of maize processing wastewater by a five-stage reactor cascade resulted in increased gas production rates and improved quality gases (up to 91% CH$_4$) when compared to reactor configurations with fewer stages [33]. At an HRT of 23 days, 90% DOC removal was recorded in the first reactor of the cascade, but removal efficiency at later stages was negligible, amounting to less than 5%. However, when the HRTs were reduced, less than 80% DOC removal occurred in the first stage, but almost 10% was recorded in the second. Gas production over the cascade improved at lower HRTs with increased CH$_4$/CO$_2$ ratios and the reduction in pH over the five reactors was reported to enhance CH$_4$ generation. Organic loading rates of up to 1.7 kg COD m^{-3} d^{-1} were employed.

The operating characteristics and economics of a fully-operational, full-scale two-stage anaerobic digestion system constructed to treat a daily output of 600 kg manure from 4000 caged laying hens (a waste typically high in ammonia) have been outlined [34]. The system consisted of a primary digester loaded directly by gravity flow and diluted by water at 60 °C, connected in series to a secondary digester. The effluent displaced from the first unit flowed directly into the second, and both digesters, of the packed bed design, were located partially below ground. The operating temperature of the first unit was 50 °C i.e., thermophilic, and sufficient heat was maintained in the effluent from this reactor to maintain the second unit in the mesophilic range (30–35 °C). Loads of up to 11.3 kg VS m^{-3} d^{-1} could be applied at HRTs of 5 days to the primary digester; up to 70% VS removal was recorded but the system showed signs of instability at sustained high loading rates, although temporary overloads and periods of no loading were tolerated by the digester assembly.

The major advantages in operating more than one stage in the anaerobic digestion process are the increased levels of COD removal, enhanced biogas production and decreased detention times. However, each reactor in a staged process contains bacterial populations which perform similar conversions, although the proportions of these may differ in each vessel.

9.3 Two-Phase Digestion

The two-phase anaerobic digester, structurally similar to the two-stage system (see Fig. 26) is based on the premise that the environmental conditions pertaining in most anaerobic waste digesters are not optimal for both fermentative and methanogenic microorganisms. Volatile acids appear consistently as detectable intermediates in degradation processes, prior to their conversion to stable end-products, principally CH_4. The sequential biochemical conversions occurring during digestion are attributed to discrete microbial populations that must exist symbiotically to ensure maximum system efficiency. A fragile balance exists between VFA production and utilisation and if it is upset, the sensitive methanogens may be repressed by the more rapidly-growing acid-formers to such an extent that the detrimental reactor conditions engendered cause system failure. If the biphasic reaction process can be physically separated by dialysis techniques or by the application of chemical or kinetic controls, both phases can operate under optimal conditions, with the first unit receiving the raw or previously-conditioned wastestream and the second unit the effluent from the first, with adjustment between phases if necessary [35]. To operate such a system, conditions in the primary reactor must be made sufficiently unfavourable to force the methanogens to grow only in the second (digestion being a non-sterile process with a continued inoculation, in many cases, of methanogens and other bacteria from the waste).

Hobson [14] argued that if the first step of digestion is the hydrolysis of polymers, a two-phase digester assembly is not feasible, due to the protracted detention times needed for the conversion. Massey and Pohland [35] suggested that the process could be applied to complex as well as simple solution-type substrates, and equations were derived describing the growth of bacteria during substrate-utilisation in the dual-phased treatment system. Research undertaken to provide kinetic predictions for the phase separation of the anaerobic degradation process has been reported. Estimated kinetic parameters for the acidogenic and methanogenic phases were obtained by sequential reductions in HRT in a single-staged completely mixed reactor to the point at which the production of VFAs was the dominant reaction [36]. The operation of a two-phase anaerobic process employing sludge as the primary substrate has also been recorded [37] although estimates

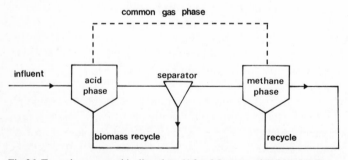

Fig. 26. Two-phase anaerobic digestion. (After Massey and Pohland [35])

of the kinetic parameters were retarded by problems in the measurement of the concentrations of active bacteria, as recycle and biomass separation facilities had not been included in the reactor system or phases.

Physical phase separation, by means of a vinyl dialysis membrane, was used to estimate the pH and oxidation/reduction potential for the acidification and CH_4 fermentation of digester sludge [38]. Under optimal conditions for the methanogenic phase, the acidification step operated at 75% of maximum and was therefore rate limiting. By adjustment of pH, temperature and residence times, optimum conditions for acidogenesis and gas formation may, however, be obtained.

In the treatment of primary sludge, two anaerobic upflow reactors (converted from conventional CSTR systems) in series were compared with a single-staged CSTR at 35 °C [39] in order to determine the optimal HRTs for the primary and secondary phases of anaerobic digestion in the system. Digestion in the dual assembly occurred at higher loading rates and lower HRTs than those recorded for the single-phase unit. Generally, the dominant conversion in the primary reactor was acid fermentation, whilst methanogenesis occurred mainly in the second vessel. Sludge was removed from the acidogenic reactor at 10% (v v^{-1}) and fed to the secondary stage. Consideration of VFA production, pH, HRT, loading rates and CH_4 yield indicated that an HRT of between 1 and 1.5 days would be optimum for first-phase digester operation, and an HRT between 4 and 5 days appeared optimum for the methanogenic stage. In a dual-phased system comprising two laboratory stirred tank fermenters however, 2–4 days retention time was necessary for the acidogenic phase of liquid swine waste digestion, while the maximum CH_4 yield of the methanogenic phase was obtained at an HRT of 12 days [40]. The limiting factors in the latter investigation were the high lignin and crude fibre contents of the waste, and kinetic constants were estimated to aid process optimisation.

The advantages of two-phase anaerobic digestion operations include enhanced CH_4 yields, reduced total reactor volume and superior effluent qualities. Ghosh et al. [39] enumerated other advantages in primary sludge treatment:

1) reduction of system capital costs relative to conventional processes;
2) operation with net production of energy;
3) substantial increase in solids reduction – the cost of digested sludge disposal was estimated at less than half that of conventional processes; and,
4) non-interference with reactor performance by the 40–60% (v v^{-1}) metal-containing industrial wastes which comprised the feed sludges.

Additional benefits of the two-phased anaerobic stabilisation system have been outlined by several workers [5, 35, 41, 42]. These include the increased stability of the system due to the more heterogeneous nature of the bacterial population and also to the buffering capacity of the acidogenic stage, the facilitated removal of acidic sludge without loss of methanogens, the rapid reactivation after shock loads of high strength (glucose-based) waste, and the potential applications of the dual-phase operation to more complex, soluble wastes.

Several conclusions have been drawn from kinetic and experimental analyses of separated-phase operations in stirred tank and upflow reactors [35, 39]:

1) by appropriate management of HRTs the acidogenic and methanogenic phases of anaerobic digestion may be separated;
2) volatile organic acids and biomass are the main products generated by the catabolism of substrates in the acid phase reactor, whereas CH_4 production is minimal;
3) acetate is the principal VFA formed;
4) system operation can be practically improved by the introduction of a gravity sedimentation and recycle stage before the application of the effluent to the methanogenic reactor, and by the incorporation of a common gas phase between the two reactors;
5) gravity sedimentation followed by recycle does not appear to be practical at the methanogenic stage due to poor settleability of reactor solids;
6) kinetic analyses of HRT data indicate that maximum specific growth rates of $2.7\,h^{-1}$ for the acidogenic bacteria, and $0.018\,h^{-1}$ for the methanogens are possible; and,
7) simple substrates can be converted to end-products within about one day in the acid phase and 15–20 days in the methanogenic phase.

The poor settleability of the anaerobic biomass from the secondary reactor unit in suspended-growth phased systems precludes the use of sedimentation and recycle as means whereby system efficiency may be improved. However, concentration of biomass by culture enrichment or by the provision of attachment media or surfaces may overcome the problem. A reactor configuration such as the anaerobic fluidised bed, that is capable of high biomass retention by particle-surface immobilisation or entrapment, can provide such a solution.

In a comparison of a single-staged anaerobic fluidised bed with a two-phased system consisting of a stirred tank acidogenic reactor and a fluidised bed methanogenic reactor, Bull et al. [42] reported that the separated-phase system produced an effluent with a significantly reduced SS concentration, and that under shock-load conditions, the dual-phased system was inherently more stable. In a two-phased assembly where both reactors were of the fluidised bed configuration, a 22% reduction in total reactor area and bed volume and a 34% reduction in system power requirements were reported [7]. The latter improvement resulted from the smaller reactor area necessary and also a reduction in the hydraulic loading rate to the second stage reactor. In addition, the low biomass yield and the biofilm characteristics of the methanogenic bacteria allowed the use of smaller-sized sand media in the secondary reactor, and thus a reduction of the hydraulic loading rate required for fluidisation of the particles.

The physical separation of acidogenic and methanogenic bacteria in an anaerobic digestion system utilising a 1% solution of glucose resulted in the almost total conversion of substrate to biomass and gases [43]. The acidogenic CSTR could be operated with an HRT of 10 h on such a substrate (cf. kinetic estimations above) and at 30 °C and pH 6.0 the main reaction products were acetate, butyrate, CO_2 and H_2, the latter representing a COD reduction of 12%. Lactate was only recorded during sharp increases of glucose loading rates and did not accumulate under steady-state conditions. The methanogenic reactor at 30 °C, pH 7.8 and with an HRT of 100 h converted 98% of the organic influent to biomass and biogas. Fatty acid breakdown was initiated a few hours subsequent to input and the conversion of these VFAs in terms of methanogenic cell mass was considered low, results in agreement with those of Sutton and Li [7].

The use of specific pore-sized ceramic media to immobilise microorganisms in a dual-phased digestion system has also been described [8, 10]. The first reactor was operated as an hydrolytic-redox phase, contained predominantly facultative acidogenic bacteria and was operated at a temperature higher (30 °C) than the second, methanogenic phase. The primary phase reactor employed an upward flow delivery with no recycle facility and was assembled vertically under pressure. The waste molecules (cellulose, proteins) were reduced by both hydrolytic and redox mechanisms, any oxygen in the system was removed, and CO_2 was produced. Due to the vertical assembly under pressure, large quantities of CO_2 were dissolved and transferred in the effluent to the horizontally-assembled methanogenic phase. The second phase was found to contain predominantly *Methanobacter* spp. As this phase was also under pressure, high concentrations of CO_2 remained dissolved and were thus reduced to CH_4, an hydrophobic gas that was easily liberated. When sewage adjusted to pH 8.6–8.9 and containing 800–2,600 mg COD l^{-1} was continuously delivered to the reactor system at HRTs of 2–5 h, the COD was reduced by 70–88%. Approximately 45% of the total carbon influent was converted to CH_4 and the effluent biogas contained greater than 90% CH_4 with less than 5% CO_2. The efficient performance of the two-phase system was attributed to the optimised pore dimensions of the media, which maximised cell accumulation and prevented washout, the increased conversion of CO_2 due to the increase of pH above 8.0 and the pressurisation of the reactor assembly, and the maximisation of the gas/fluid interface for removal of CH_4 by the horizontal configuration of the methanogenic reactor. In addition, the lower temperature of the second stage contributed to CO_2 retention in the fluid phase. The immobilised bacterial population was not adversely affected by shock-loads of feedstock, nor alterations in influent composition and reactor performance recovery times were 1–2 days. It was also observed that the porous ceramic media containing the *Methanobacter* could be removed from the reactor stage and exposed to the air for several hours, with complete recovery occurring within 12 h after its replacement. The system was used to treat a variety wastes including manures, sulphite-pulp-wastes, pumpkin wastes, whey and stillage of 1–60 g COD l^{-1}, with promising results.

Two pilot scale anaerobic fluidised bed reactors containing 0.1–0.3 mm diameter sand particles as media were used firstly for the acidification of baker's yeast production wastewater and secondly to convert the acidified effluent to CH_4, each process confined to one reactor [5]. The primary phase reactor gave complete acidification within 1 day after start-up; sulphate-reducing activity and CH_4 generation were both initiated at the 4 week stage. In the second reactor the biomass concentration reached 40 kg VSS m^{-3} giving conversion rates of 50 kg COD m^{-3} d^{-1} at HRTs of 1 h and biogas production of 0.18 m^3 kg^{-1} COD removed. On the basis of these results, the design and construction of a full-scale anaerobic fluidised bed treatment system consisting of acidogenic and methanogenic reactor beds each of about 300 m^3 volume, was undertaken. Treatment of approximately 200 m^3 h^{-1} of wastewater has been estimated, resulting in the removal of about 15 t COD d^{-1}.

Prospects for the phased anaerobic treatment of wastewaters are extremely promising. With the variety of reactor designs available and the amenability of reactors to modification, existing treatment systems may be replaced or upgraded as required to achieve increased stability, higher loading capacities and greater process efficiencies than are possible using single-stage systems.

References

1. Mosey F (1983) Water Sci Technol 15:95
2. Podruzny MF, van den Berg L (1984) Biotechnol Bioeng 26:392
3. van Kleef BHA (1982) Antonie van Leeuwenhoek 18:521
4. Kennedy KJ, Muzar M (1984) Biotechnol Bioeng 26:627
5. Heijnen JJ (1983) Development of a high rate fluidised bed biogas reactor. In: Proc Eur Symp Nov 1983, Noorwijkerhaout, Netherlands, p 259
6. Jeris JS (1983) Water Sci Technol 15:167
7. Sutton PM, Li A (1983) Water Sci Technol 15:333
8. Messing RA (1983) Bioenergy production and pollution control with immobilized microbes. In: Tsao GT (ed) Ann reports on fermentation processes, vol 6. Academic Press, New York London, p 23
9. Young JC, Dahab MF (1983) Water Sci Technol 15:396
10. Messing RA, Stineman TL (1983) Ann NY Acad Sci 413:501
11. Wilkie A, Colleran E (1984) Biotechnol Lett 6:735
12. Verstraete W (1983) Biomethanation of wastes: perspectives and potentials. In: Biotech 83: Proc Int Conf on Commercial Applications and Implications of Biotechnol. Online, London, p 725
13. Byrd JF (1961) Combined treatment. In: Proc 16th Ind Waste Conf, Purdue Univ, Lafayette, Indiana 1961. Ann Arbor Science, Ann Arbor Michigan, p 92
14. Hobson PN (1982) Production of biogas from agricultural wastes. In: Subba Rao NS (ed) Advances in agricultural microbiology. Butterworth Scientific, London, p 523
15. van Assche P, Poels J, Verstraete W (1983) Biotechnol Lett 5:749
16. McConville T, Maier WJ (1979) Biotechnol Bioeng Symp 8:345
17. Rudolfs W, Trubnick EH (1935) Sewage Wks J 7:852
18. Adams AD (1975) Water Sewage Wks 122:46
19. Adams AD (1975) Water Sewage Wks 122:78
20. Spencer RR (1979) Biotechnol Bioeng Symp 8:257
21. I.C.I. United States Inc. (1977) Improved anaerobic digester performances with powdered activated carbon, report no 903-4, I.C.I. US Inc
22. Henze M, Harremoes P (1983) Water Sci Technol 15:1
23. Melcer H, Nutt S, Marvan I, Sutton P (1984) J Water Pollut Control Fed 56:191
24. Nutt SG, Melcer H, Pries JH (1984) J Water Pollut Control Fed 56:851
25. Pretorius WA (1971) Water Res 5:681
26. Pretorius WA (1973) Anaerobic digestion of raw sewage. In: Adv in Water Pollut. Res.: Proc 6th Int Conf. Pergamon, Oxford, p 685
27. Hashimoto S, Fujita M, Terai K (1982) Biotechnol Bioeng 24:1789
28. Suidan MT, Strubler CE, Kao S-W, Pfeffer JT (1983) J Water Pollut Control Fed 55:1263
29. Cross WH, Chian ESK, Pohland FG, Harper S, Kharkar S, Cheng SS, Lu F (1983) Biotechnol Bioeng Symp 12:349
30. Harper SR, Cross WH, Pohland FG, Chian ESK (1984) Biotechnol Bioeng Symp 13:401
31. Khan KA, Suidan MT, Cross WH (1981) J Water Pollut Control Fed 53:1519
32. Cho YK (1983) Biotechnol Lett 5:555
33. Civit E, Durán de Bazúa C, Gonzalez S, Hartmann L (1984) Environ Technol Lett 5:89
34. Steinsberger SC, Shih JCH (1984) Biotechnol Bioeng 26:537
35. Massey MC, Pohland FG (1978) J Water Pollut Control Fed 50:2204
36. Ghosh S, Pohland FG (1974) J Water Pollut Control Fed 46:748

37. Gosh S, Conrad JR, Klass DL (1975) J Water Pollut Control Fed 47:30
38. Hammer MS, Borchardt JA (1966) J San Eng Div Proc Amer Soc Civil Engr 95(SA5):907
39. Ghosh S, Sajjad A, Henry MP, Bleackney RA (1984) Biotechnol Bioeng Symp 13:351
40. Cseh T, Czaco L, Toth J, Tengeroy RP (1984) Biotechnol Bioeng 26:1425
41. Cohen A, Zoetemeyer RJ, van Deursen A, van Andel JG (1979) Water Res 13:571
42. Bull MA, Sterritt RM, Lester JN (1984) Biotechnol Bioeng 26:1054
43. Cohen A, Breure AM, van Andel JG, van Deursen A (1982) Water Res 16:449

10 Start-Up of Anaerobic Bioreactors

One of the few major problems that recurs in the application of anaerobic bio-treatment to wastewater is the frequent recalcitrance of start-up procedures. It is widely observed in the literature that a significant amount of down-time is involved in the initial start-up of most, if not all, anaerobic reactor systems. The main difficulty appears to be the development of the most suitable microbial culture for the wastestreams in question. Once the biomass has been established, either as a granular particle or floc system, or attached to inert carrier media as a biofilm, the operation of the reactor is generally quite stable. The sensitive nature of the majority of anaerobic bacteria and the extreme oxygen lability of the enzyme systems of obligate anaerobes render the reactor population more susceptible to slight fluctuations than is the case under comparable aerobic conditions; start-up of an anaerobic system is consequently more time consuming than initiation of an aerobic process.

10.1 Nutrient Balance and Inhibition at Start-Up

Although a specific quantity of nutrient is required for the growth and maintenance of the cell, the effect of excessive levels of nutrient on start-up of the digestion process must also be taken into consideration. Excess concentrations of many nutrients are inhibitory rather than stimulatory to anaerobic degradation. A micronutrient may interfere specifically with an enzyme system or membrane component; non-specific inhibition may be the result of osmotic effects by macronutrients [1]. In addition, effects on microbial metabolism may occur at the genetic level of transcription, resulting in catabolic repression. The conversion of substrate may be substantially altered by repression of catabolic enzymes, although growth is in general rarely affected by repressive phenomena [1]; carbon, sulphur, phosphate and nitrogen are particular effectors of this species of inhibition. The inhibitory effects mediated by several nutrient sources when present in excess necessitates the careful regulation of substrate supply to the reactor during start-up. Nutrient loading should be a gradual process, increase of supply level keeping pace as far as possible with increase of biomass, until the steady state is reached.

The adequacy of the carbon source is crucial for the optimal growth and activity of the reactor flora. In common with many other nutrients, however, an increase of carbon source concentration will reach a limit above which the microbial growth rate decreases. This effect is known as substrate inhibition (cf. catabolite repression, above, which rarely affects growth rate, although the carbon source can also have this effect). The typical level of substrate inhibition by carbohydrates operates at about $100\text{--}150 \text{ g} \text{l}^{-1}$ [1]. Osmotic pressure is frequently the cause of this effect: an increased external concentration of substrate partially dehydrates the microbial cell and results in a reduction of growth rate. Carbon substrates such as phenol, butanol or toluene exert more specific effects, causing damage to cell membranes by extraction of cell components. These substances are inhibitory at concentrations as low as 2 or $3 \text{ g} \text{l}^{-1}$. High throughput and recycle rates in some reactor designs ameliorate these effects considerably, and efficient breakdown of recalcitrant organic compounds can occur [2–6]. The carbon source must therefore be maintained at a sub-critical level to ensure efficient conversion rates and biomass growth.

10.2 Seeding and Loading Regimes

Anaerobic microorganisms exhibit growth rates which are much slower than those of aerobes; adequate seeding of the anaerobic reactor at start-up is thus more critical than is the case for aerobic systems. In fixed-film systems, the quantity of seed utilised should be at least 10% of the reactor volume [5] otherwise start-up may fail or build-up to steady-state may be prolonged for at least a year. A seed inoculum of 30–50% reduces the start-up time required to a considerable extent [5].

In suspended-growth biomass systems such as the UASB and contact processes, where the potential for wastewater treatment is primarily dictated by the dual parameters of the quantity of biomass retainable in the vessel and the specific activity of the retained biomass, an inoculum of about 30% reactor volume of active sludge is required. The activity of the initiating biomass inoculum affects the rate of start-up, as does the type of feedstock utilised. In anaerobic filter (packed bed) reactors, the use of seed sludge from an identical process to initiate the digestion system was reported to increase the rate of start-up fourfold, in comparison to the rate achieved when seeding with a general municipal digester sludge [6]. In one investigation, the more active the initial seeding dose proved to be (i.e. the higher its content of active SS), and the more complex the waste it had been grown on, the more readily were the digesters in the systems started up [7]. The reasons for these increases in start-up rate in the above systems were related to a great extent to the microbial populations initially present in the seeding material. In the former case, the bacteria required to degrade the substrate were already present in the seed sludge utilised; with the general municipal sludge, the relevant populations had to be selected for by the environmental conditions, i.e., the substrate, before conversion to product could effectively occur. Where the seed sludge had been grown up on a complex substrate, the presence of a hetero-

150

geneous microbial population ensured rapid selection of the relevant organisms upon introduction into the digester system.

Many wastestreams are deficient in certain microbial nutrient sources: domestic sewage, for example, is not particularly rich in carbon, but has a high nitrogen concentration. If the feed to be treated is deficient in some essential nutrient, it should be supplemented during the start-up phase with the required substances, or with an additional wastestream of feedstock which is rich in the nutrients lacking in the primary waste. A period of acclimation to the waste load may be necessary before an increase of COD loading rate can be attempted. It has been indicated that anaerobic sludge which had been fed on sugar-beet waste for a period adapted rapidly to a substrate consisting of a mixture of alcohols. The sugar-beet-fed anaerobic sludge was in fact acclimated to the VFA content of its feedstock, and was therefore suitable seed material for the start-up of a reactor to be utilised in the processing of alcohol-containing wastes. It was reported that 2–4 days sufficed for the initiation of methanol degradation, while the conversion of higher alcohols was observed to start almost immediately. Overall, however, 4–8 weeks were necessary for the start-up of the UASB and the attainment of a steady-state condition [8].

If seed sludge from a fully-operational digester treating waste similar to the feedstock in use is unavailable, a general inoculum of municipal digester sludge or manure can be used. Digested sludge from a domestic sewage treatment operation generally provides an adequate seed, but supplementation with a carbon source and essential nutrients may be necessary initially and frequent reseeding of the reactor is recommended throughout the start-up phase. The provision of trace elements such as metals during this initial period has also been reported to improve operation [5, 8, 9]. These trace elements may aid digestion by supplying necessary cofactors for the metallo-enzymes required in catabolism, or by maintaining membrane or cell-component stability. Acclimation to a novel waste substrate may result in as much as one or two months down-time before operational reactor status is achieved; efficiency even then may be less than maximum.

Methanogenic bacteria have relatively long generation times of 0.5–2.0 days and consequently give a poor sludge yield: 4–8 months may be required for the attainment of a microbial steady-state in suspended biomass systems, which are more subject to biomass washout than their fixed-film counterparts. In a thermophilic operation, with an even smaller sludge yield, considerably longer start-up periods – in some cases exceeding a year – may be necessary [10]. Using an expanded bed, fixed-film thermophilic bioreactor system, however, steady-state conditions after 5 months of operation were reported by Schraa and Jewell [11], and less than 3 months were necessary for the attainment of steady-state in the anaerobic mesophilic fluidised bed system of Rudd et al. [12]. In contrast, the conventional digester with a recycle facility utilised by Rimkus and others [13] evinced erratic VFA levels for the first 18 months of operation. At start-up of the above expanded bed system, the reactors had been inoculated with active populations of both thermophilic and mesophilic organisms; the thermophiles originated from a conventional thermophilic laboratory anaerobic digester fed with animal waste and the mesophiles from the digester sludge of a local wastewater treatment plant. The organisms necessary for the thermophilic expanded bed were thus probably present as a portion of the inoculum. However, Rimkus et al. [13] raised the temperature of a mesophilic digester by 30 °C at a rate of 0.6 °C d^{-1} at regular intervals of 2–3 weeks to allow for digester stabilisation: a 14 day detention time was also operated during temperature elevation. Subsequent to start-up and the attainment of a mesophilic steady-state in 50 days, the temperature of the fluidised beds of Rudd et al. [12] was increased from 37 to 57 °C by 1 °C d^{-1} over a further period of 20 days, and the thermophilic steady-state achieved after 13 days of stabilisation. The selection of thermophiles from a mesophilic population is reported to require a long acclimation period: Chin [14] found that only 9% of a mesophilic digester flora were capable of growth at elevated temperatures (55 °C). Start-up of a thermophilic operation by the above procedure is therefore prolonged and must be carefully monitored to prevent souring of the digester by excess VFA production.

The development of a granular active sludge is necessary for the start-up of UASB reactors, as sludge retention is primarily a function of sludge settleability. In an investigation into the acclimation of digester sewage sludge during start-up of UASB reactors, sludge of a high activity could be developed within a 6 week period using acetate and propionate as feed [15]. These VFAs were chosen because acetate is the precursor of about 75% of the CH_4 produced from the anaerobic digestion of the majority of complex substrates, and propionic acid, an important intermediate in the degradation process, is frequently the cause of inhibition of methanogenesis during start-up. As divalent cations are known to aid flocculation of anaerobic sludge and thus enhance retention within the reactor, the above investigators added varying concentrations of calcium ions (Ca^{2+}) to their UASB system; positive effects were observed during the course of the experiment.

Raw primary and secondary sludges were utilised to generate active seeding sludges by Schwartz et al. [16]. These materials were digested at 20 day retention times in small 1 l anaerobic digesters, at 35 and 50 °C, and the sludges developed were subsequently used to seed upflow anaerobic filter reactors. The 35 °C seeding digesters were operated for 2 months and the thermophilic reactors for 3 months, the raw sludges being replaced gradually by heat-treated sludge and decant liquor for meso- and thermophilic reactors respectively. A seeding load of 10% of the anaerobic filter reactor volume was used to initiate start-up. The initial HRT exceeded 3 days while the feed was adjusted to pH 7.0 and diluted 1:1 with water to prevent toxic shock to the system. Full strength liquor feed, with no pH adjustment, was reported to be tolerated by the system within a few weeks and a stable population of anaerobes had developed by the sixth week. The importance of avoiding drastic fluctuations in the influent, and consequently the environment of the bacterial flora during the critical establishment phases of microbial development, is paramount. In fixed-film systems such as the anaerobic filter, initial biofilm attachment to the carrier media is the primary and perhaps most difficult aspect of start-up; pH shocks in particular can alter the charge balance of the system, thus interfering with adhesive and cohesive binding forces.

Rotating biological contactors (RBCs) are also biofilm-reliant systems. During the start up of an anaerobic RBC process, methanol was used to augment a sucrose feed [17]; the alcohol was utilised to encourage the establishment of a

152

methanogenic microbial subpopulation. A batch system was operated for the first 6 days, with periodic recycling of settled effluent solids. The feeding pattern was then altered to a continuous cycle, but effluent solids recycle remained periodic. The start-up of this system was considered successful on day 27, although only a quasi-steady-state had been achieved.

Considerable variations are found in the lengths of time necessary for the start-up of different systems. In another fixed-film reactor also of the anaerobic filter or packed bed configuration, which had been designed to deal with raw municipal wastewater at ambient temperatures (10–25 °C), the seeding sludge employed originated from an anaerobic digester which had previously been used to treat combined sludge from a municipal sewage treatment system [18]. The average solids concentration of the seed sludge was adjusted by dilution with primary wastewater to approximately $6\,000\;\mathrm{mg\,l^{-1}}$, after which the mixture was circulated throughout the bioreactor in a closed circuit for 3 days, to permit distribution of the solids through the system. Cessation of the cycle allowed the solids to settle on the packing. The experiment did not utilise a synthetic substrate to enhance start-up, but commenced feeding with primary wastewater on a continuous basis 48 h after the arrest of recirculation; increase of feed rate was based on the filters' removal of soluble COD and on gas production. An active methanogenic population in the reactor was indicated by an increasing generation of CH_4 while the presence of acidogenic bacteria was signified by high total VFAs; influent flow rates over 60 days were increased linearly by ten-fold. This regime, however, resulted in a drop in effluent pH due to an accumulation of VFAs; the acidogenic bacteria were evidently progressing more rapidly than the methanogens, which have a slower growth rate in general than the majority of acid-producing bacteria (the OHPA bacteria being an exception to this). Loading and HRT were decreased to accommodate the pH fluctuation, and start-up overall required 9 months to complete.

An appraisal of the start-up of anaerobic fluidised beds was made by Bull et al. [19] using four regimes: two of the reactors were loaded continuously at 4.6 kg $COD\;m^{-3}\,d^{-1}$, and two were step-loaded over a period of 40 days; methanol replaced 50% of the COD load in one continuously-loaded and one step-loaded system. In all the systems, COD reductions in excess of 90% were attained at loadings of $4.8\;\mathrm{kg\,m^{-3}\,d^{-1}}$ within 40 days. The most efficient start-up regime was considered to be the stepped-loading process with initial addition of methanol as a substrate to enhance the growth of methanogenic bacteria; the absence of operational difficulties using this system indicated that markedly greater COD loadings could be applied, with resultant rapid start-up and continued reactor efficiency.

10.3 Washout

The start-up of non-attached biomass processes and those systems which rely on suspended as well as attached microflora, is prone to washout of biomass during the initial phase of the procedure. Experiments with the UASB reactor in particular reveal significant losses of active microbial mass during this stage; the amount of seed sludge retained in the reactor vessel is small and hence overall bioactivity is low. Washout of seed sludge returned to the UASB by recycle has been noted to recur within a few days [20]. The specific activity of the sludge displaced from the reactor under such circumstances has been found to be similar to that of the sludge retained, i.e., a portion of the net microbial growth occurring during the initial phases of start-up was also removed with the effluent.

The fraction of the sludge remaining within the reactor has superior settleability and the generation of active microbial mass occurs in and on the settled solids. The specific activity of the retained bulk hence increases rapidly if the amount of sludge within the reactor is adequate for process operation. Unlimited biomass washout obviously cannot be tolerated: as the percentage of retained sludge decreases, the period of quiescence prior to the increase in biogas generation lengthens and the approach to steady-state is prolonged. The initial retention of biomass, therefore, is critical to reactor activity during the start-up phase. The minimisation of biomass washout can be achieved by the utilisation of a suitable type of seed sludge. Evidence indicates that a thick, relatively inactive seeding mixture tends to promote enhanced retention of the sludge within the reactor vessel. A VFA mixture (equal proportions of acetate, propionate and butyrate) was treated in a 30 l UASB reactor which employed a concentrated digested sewage sludge (40 g VSS l^{-1} and 47% ash) of low specific activity (0.4 kg VSS d^{-1}); a relatively efficient sludge retention time with a limited sludge concentration of 4.9 g VSS l^{-1} was reported [15].

Optimal growth conditions within the reactor can be assured by alteration of the wastewater composition when necessary; the latter can be achieved by the addition of nutrients, vitamins or flocculation-enhancing agents to the influent stream. The increased settleablity of sludge in one UASB reactor was attributed to a limited extent to the precipitation of calcium carbonates and phosphates from the hard water used in the reactor system: it was suggested that the precipitated salts "weighted" the sludge pellets, hence increasing their sedimentation capabilities [21].

Careful reactor design and process management can also aid the minimisation of biomass loss during start-up. The gas/solids separator device, incorporated usually internally in the UASB system, is composed of a series of angled baffle plates and/or a cone-shaped wall at the top of the reactor. The angle of incline of the solids separator wall and the size of the gas venting orifice will affect the washout rate. The 50 ° wall angle of the settler used in the reactors of Lettinga [15] appeared superior to the 45 ° wall-angled design of Schwartz et al. [16]. During the start-up phase, extended periods of over- and under-loading and fluctuations in temperature, pH, HRT and recycle ratio must be avoided. Overloading affects start-up adversely because the gas production in the gas/solids separator hampers sludge settlement by causing gas-lift in the fermenter and flotation of the sludge. Underloading results in the formation of a massive, compacting sludge [20].

The loading rate in the initial phase of start-up of an anaerobic reactor must be low e.g. around 0.1 kg COD kg^{-1} VSS d^{-1} which corresponds to 1.2 kg COD

154

$m^{-3} d^{-1}$ for a system with a concentration of biomass of 10–20 kg VSS m^{-3}. When an increase in gas production is observed, the loading rate can be increased stepwise, by up to about 50% per week depending upon the type of reactor employed and the wastestream under treatment.

One investigation revealed that when the seeding load of a suspended growth (contact) digester had acclimated to the waste, the rate of organic loading could be increased relatively rapidly and steadily by a factor of 2 in 20 days [7]. However, fixed-film (anaerobic filter) reactors could not tolerate a consistent doubling of the COD loading rate and performance was dependent on waste type and reactor size. Thirty days or more were necessary to allow a doubling of the load rate when simulated sewage sludge was employed as a feedstock; with bean blanching waste, load rate could be doubled in 10–15 days, with 0.8 l volume vessels, whereas 40–50 days were necessary to reach twice this rate when 35 l reactors were utilised. These results reflect the variability encountered in the degree of mixing in fixed-biofilm reactors during initial operation.

The start-up of fixed-film systems such as expanded and fluidised beds and anaerobic filter reactors appears to depend to a major extent on the initial establishment of the biofilm. The film support material has been found to affect both the rate of the development and the ultimate maximum performance of the stabilised film [7]; porous inert support materials tend to promote biofilm growth. In an experiment devised to test the effect of different support materials on the establishment of a biofilm and the activity of its associated flora, it was observed that the film developed much more rapidly on needle-punched polyester supports and on red draintile clay media than it did on potter's clay and polyvinyl chloride (PVC) carrier materials. A substantial lag was recorded prior to film development on the PVC media, although the substance was not in itself inhibitory to the establishment of methanogenic bacteria [22]. On all four media tested cessation of the production of biofilm was abrupt when the maximum steady-state performance was attained. It was also reported that film development was frequently halted by the occurrence of a pseudo-steady-state phase lasting up to 100 days in the bioreactors, especially in those containing potter's clay. This phenomenon had been observed previously during the examination of the effect of load increases in fixed-film reactors [7]. Large increases in loading rates were tolerated by the biofilm-coated red draintile clay media and the film was observed to produce larger volumes of CH_4 than those on the other support materials, although rate of film development did not relate directly to steady-state performance. The highest rates of film formation were similar to the maximum growth rates of methanogenic bacteria. The characteristics of the carrier media that enhanced film development were deduced to be the ease of entrapment and attachment of bacteria to the materials, and also the probable leaching of minerals from the red tile clay.

Another investigation into the efficiency of start-up of anaerobic filter reactors using different carrier media was undertaken by Wilkie and Colleran [23], who evaluated clay, coral, mussel shell and plastic supports. The most rapid increase of COD removal from the system was attained by the use of clay carriers which were of a 69% porosity and which reached 70% COD reduction in 20 days. The plastic media, of porosity value 94%, required 50 days for a similar removal of COD; all the reactors employed were maintained at a COD loading of 5 kg

COD m^{-3} d^{-1}. The clay-packed system also supported a slightly superior CH_4 yield over the other reactors; the mussel shell proved to be the least efficient carrier for COD reduction, CH_4 production and rapidity of start-up, the latter requiring 53 days for completion. The clay support also presented the lowest surface to volume ratio of the media evaluated, being four times less than that of the coral. The start-up performances achieved were not therefore considered to be functions of either unit surface area of carriers or their relative porosity, results in contrast to those of Kennedy and Droste [24], who reported that start-up and performance of downflow stationary fixed film (DSFF) reactors were enhanced by increased surface area. Support materials such as clays, which leach minerals, may prove more advantageous in the start-up of fixed-film systems by supplying some nutrients most essential to attachment processes.

Poor adhesion of the seeding biomass to carrier particles has posed problems in anaerobic biofilm reactors. For the vigorous attachment of microorganisms to supporting matter, the surface of the carrier particle should ideally be porous and the outer layers of the attaching cells sticky or in some way adhesive. Microorganisms that do not generate extracellular polymeric material adhere to the slime matrix produced by other bacteria in the population. Micrographs of both thin and thick biofilms have shown that the former tend to consist of a layer of bacteria on the support particles, whilst the latter show a similar density of organisms per unit surface area, but with a much thicker slime matrix [5].

In aerobic bioreactors, biofilm attachment to media is rarely problematical as aerobic microorganisms continuously produce slime in response to a carbonaceous waste, the most advantageous substrates for extracellular polymer production being carbohydrates, especially the oligo- and polysaccharides. Under anaerobic conditions, however, extracellular material is infrequently produced and attachment therefore, especially to non-porous materials such as sand, may cause problems. The addition of cane sugar or dextran to feedstock at start-up has enhanced attachment, as has the inclusion of slime producing organisms and their substrates (e.g. *Leuconostoc* spp. and sucrose). Bull et al. [19] manipulated the C:N ratio by the addition of methanol to enhance microbial production of extracellular polysaccharide and hence encourage adhesion. The addition of polyacrylamides and other cationic polymers has been employed to increase the flocculation characteristics of microorganisms [25]; surface-roughened particles will also tend to aid entrapment.

One explanation of the poor performances achieved in the early phases of start-up in fixed-film reactors may be that the initial populations of organisms attached to particle surfaces in anaerobic digesters have not developed the symbiotic relationships necessary for the efficient operation of many of the conversion pathways required during anaerobic degradation processes. Acclimation speed of populations to the prevailing environmental conditions within a digester necessarily includes the development of these interdependent associations, and to allow these microstructures to stabilise during start-up, the reactor must be protected from fluctuations in temperature, pH, nutrient concentration, recycle rate and HRT. Close monitoring of digester performance through VFA production and gas evolution is particularly required during the start-up period, and load increases should be halted immediately any retrogression of performance is ob-

156

served. During the start-up of anaerobic fixed-film reactors, the volume of influent feed in relation to biomass should be maintained at a low level, with microbial density in excess of 20 kg VSS m^{-3} (reseeding as required) and organic load less than 0.1 kg COD kg^{-1} VSS d^{-1} [5]. An HRT of more than one day is necessary initially, while retention times of up to 20 days may be advantageous. As calcium ions may aid the adhesion of microorganisms to the media, the use of $Ca(OH)_2$ rather than NaOH for the maintenance of alkalinity has been recommended [5]. In a dual-phased system, where acidogenic and methanogenic processes are separated physically, start-up of the methanogenic reactor is enhanced by utilising an acidified waste as feed.

Carrier-assisted reactor systems, especially those of the expanded and fluidised bed configuration, do not generally experience problems of washout. The support particles, being of small surface area to volume ratios, are dense enough to retain the active biomass, once attached, within the body of the reactor. The fragile balance between loading and loss of biomass that is so necessarily a part of suspended-growth processes is less important in fixed-film systems, although some consideration to influent load and seed inoculum volume must be given. According to Heijnen [26], the fluidised bed system can be started up with minimum attention to loading or conversion, the primary factor in stable operation being attachment by pressures of selection, the bacteria being forced either to adhere to the media or wash out. However, complete development of methanogenic activity was reported to take around 6 months, whereas the four start-up regimes of Bull et al. [19], entailing careful consideration of influent composition and loading procedures, were complete within 40 days. Nine months have been reported for full development of the film in expanded bed reactors [27]. The rapid start-up of a fluidised bed reactor has been described in which bed seeding was initiated by the throughput of a mixed feed of settled sewage at the flow velocity selected for the experimental regime [28]. The growth of the biofilm was evident after 4–5 days of operation. Start-up of the system was also successfully achieved as a batch process using total recycle with sand as support and a seeding mixture of settled sewage and nitrified effluent in equal proportions. The system was switched to partial recycle after 4 days of operation and the recycle ratio reduced as the biofilm on the support particles developed.

Contrasts and similarities are evident in the start-up regimes of anaerobic fixed-biofilm and non-attached biomass processes. The volume of seeding material in attached growth reactors should typically comprise at least 10% of the reactor volume; in contact and sludge blanket operations, the quantities of seed sludge necessary to adequate reactor start-up need to be much greater, of up to 50 kg dry weight per m^3 of total volume. The sludge must also evince good settling properties, without the tendency to bulk. The UASB system is loaded incrementally in order that a conversion level of around 80% is maintained; this type of regime results in approximately two thirds of the biomass washing out during the initial 6 weeks, after which time the microstructural granules or pellets develop. The total time of start-up in fluidised bed systems is determined to some extent by the rapidity and strength of adhesion of the microorganisms to the support media and also by the organisms' growth rates within the structured biofilm matrix. High fluidisation velocities of up to 10–15 m h^{-1} are applied and HRT

157

may be of the order of 1–2 h once the film has developed; recycle is necessary at 2–20 times the feedrate for film establishment. Both attached-film systems and suspended growth processes may typically require 6 months for complete methanogenic activity to become established, although individual regimes can alter this period markedly. Process parameters must be maintained within the confines of anaerobic microbiological tolerance limits in both cases and supplementation with nutrients and trace essential elements enhances both types of system performance.

10.4 Turbulence and Shear

The physical forces attendant upon reactor operation during the initation of an anaerobic system arise from the movements of particles in aqueous media (hydrodynamic forces), are externally generated (hydrodynamic shear, gravitational forces) or are the consequence of the mutual proximity of media (electrodynamic and electrostatic forces).

Where external shear fields are absent, the presence of hydrodynamic forces mediate factors such as particle diffusion, aggregation and deposition. Where no external energy input to the system occurs, particle-particle interactions, and also particle-substrate interactions, are not affected by hydrodynamic forces. However, in all biological situations external forces operate.

When two smooth surfaces are brought together with a wetting layer of liquid between them (i.e. a layer with a low angle of contact to the surface), attempts to separate them will be strongly resisted, although the surfaces have little resistance to shear, i.e., they will slide across one another with ease. A biofilm matrix tends to behave as a viscous fluid layer and the viscosity of the matrix between two adjoining surfaces promotes resistance to the shear effect in proportion to the degree of viscosity. Hence, once the film is formed, it is more resistant to the effects of shear force than flocculated or free biomass and system stability is enhanced.

The rates of transport of bacterial cells in an aqueous system under quiescent or fluid flow regimes is little understood and is subject to conjecture. The status of fluid flow (zero, laminar or turbulent) must significantly influence rates of cell transport in the early phases of microstructure development; the primary agents of transport under various environmental conditions must be determined if floc and biofilm formation within anaerobic digestion systems is to be clarified. The growth rates of microorganisms within the biofilm matrix at later stages of film evolution are influenced by transport processes such as mass transfer and diffusion of nutrients, but the roles of these during the initiation of film development are unclear.

The shear forces produced by laminar or turbulent flow regimes ensure that particle-particle approaches are close-range and rapid; for the majority of bacterial cells with diameters greater than 1 μm, moderate shear stresses in the system such as those resulting from gentle stirring or convection currents considerably enhance the frequency of collision and aid the formation of pellets or flocs. The temporary action of Brownian motion may be necessary however, to counteract any shear-induced motion during the formation of an adhesive attachment [29]. Contact and recycled bed reactors require mixing of some form to maintain the bed and flocs in suspension in the system. Agitation that produces a greater degree of interrelationship between wastewater and microorganisms may in these cases be advantageous. During the start-up phase, however, when delicate microbiological structures are being formed, over-intensive agitation may cause their dispersion.

Gas bubbles produced in the system, as has been noted elsewhere, adhere to flocs and bed particles causing them to move upwards in the reactor and wash out. Gas entrapped in the media-spaces in fixed beds can result in clogging and moving bubbles may cause channelling and short-circuiting. The release of gas bubbles from the fluid phase is retarded by agitation such that the gas hold-up time is increased; coalescence of bubbles is likewise prevented by agitation and this reduces clogging and channelling effects, especially where hydraulic flow is high in systems such as expanded and fluidised beds.

Films develop preferentially in regions of low shear [30] and dissolution of extracellular polymeric material into the bulk medium, away from the microbial cell surface, is greatest in those regions where agitation is greatest. The velocity required to expand or fluidise a bed of inert support particles therefore probably hampers the development of the biofilm. The washout of microorganisms during the early stages of adhesion and film establishment may also be exacerbated by an increased rate of biofilm detachment due to fluid shear stresses at the biofilm/ fluid interface. The biofilm may be strong enough to withstand removal pressures exerted by shear forces, or the deposition mechanism operating may prevail over the removal mechanisms of shear stress. One investigation indicated that at higher velocities, biofilm development was more dependent upon the mechanisms of deposition than upon growth within the biofilm [31] although earlier evidence supported the view that a reduction of deposition rate occurred upon increasing fluid velocity [32].

References

1. Cooney CL (1981) Growth of microorganisms. In: Rehm H-J, Reed G (eds) Biotechnology, a comprehensive treatise in 8 volumes, vol 1: Microbial fundamentals. Verlag Chemie, Weinheim-Deerfield Beach, Forida Basel, p 73
2. Evans WC (1977) Nature (London) 270:17
3. Lin Chou W, Speece RE, Siddiqi RH (1979) Biotechnol Bioeng Symp 8:391
4. Rittmann BE, McCarty PL (1980) Biotechnol Bioeng 22:2343
5. Salkinoja-Salonen MS, Nuys E-J, Sutton PM, van den Berg L, Wheatley AD (1983) Water Sci Technol 15:305
6. Young JC, McCarty PL (1969) J Water Pollut Control Fed 41:R160
7. van den Berg L, Lamb KA, Murray WD, Armstrong DW (1980) J Appl Bacteriol 48:437
8. Lettinga G, van der Geest A Th, Hobma S, van der Laan J (1979) Water Res 13:725
9. Lettinga G, de Zeeuw S, Ouborg E (1981) Water Res 15:171
10. Henze M, Harremoes P (1983) Water Sci Technol 15:1
11. Schraa G, Jewell WJ (1984) J Water Pollut Control Fed 56:226

12. Rudd T, Hicks SJ, Lester JN (1985) Environ Technol Lett 6:209
13. Rimkus RR, Ryan JM, Cook EJ (1982) J Water Pollut Control Fed 54:1447
14. Chin M (1983) Appl Environ Microbiol 45:1271
15. de Zeeuw W, Lettinga G (1980) Acclimation of digested sewage sludge during start up of an upflow anaerobic sludge blanket (UASB) reactor. In: Proc 35th Ind Waste Conf, Purdue Univ, Lafayette, Indiana 1980. Ann Arbor Science, Ann Arbor Michigan, p 39
16. Schwartz LJ, De Baere LA, Lanz RW (1982) Biotechnol Bioeng Symp 11:463
17. Tait SJ, Friedman AA (1980) J Water Pollut Control Fed 52:2257
18. Genung RK, Hancher CW, Rivera AL, Harris MT (1983) Biotechnol Bioeng Symp 12:365
19. Bull MA, Sterritt RM, Lester JN (1983) Biotechnol Lett 5:333
20. Hulsoff-Pol LW, de Zeeuw WJ, Velzeboer CTM, Lettinga G (1983) Water Sci Technol 15:291
21. Klapwijk A, Smit H, Moore A (1981) Denitrification of domestic wastewater in an upflow sludge-blanket reactor without carrier material for the biomass. In: Cooper PF, Atkinson B (eds) Fluidised bed treatment of water and wastewater. Ellis Horwood, Chichester, p 205
22. van den Berg L, Kennedy KJ (1981) Biotechnol Lett 3:165
23. Wilkie A, Colleran E (1984) Biotechnol Lett 6:735
24. Kennedy KJ, Droste RL (1984) Start-up in DSFF reactors. In: Proc 35th Ind Waste Conf, Purdue Univ, Lafayette, Indiana 1984. Ann Arbor Science, Ann Arbor Michigan, p 615
25. Treweek GP, Morgan JJ (1977) J Colloid Interface Sci 60:258
26. Heijnen JJ Development of a high rate fluidised bed biogas reactor. In: Proc Eur Symp Nov 1983, Noorwijkerhaout, Netherlands, p 259
27. Switzenbaum MS, Jewell WJ (1980) J Water Pollut Control Fed 52:1953
28. Cooper PF, Wheeldon DHV (1981) Complete treatment of sewage in a two-fluidised bed system. In: Cooper PF, Atkinson B (eds) Biological fluidised bed treatment of water and wastewater. Ellis Horwood, Chichester, p 121
29. Curtis SGM (1979) Summing up. In: Ellwood DC, Melling J, Rutter P (eds) Adhesion of micro-organisms to surfaces. Academic Press, London New York San Francisco, p 199
30. Ash SG (1979) Adhesion of microorganisms in fermentation processes. In: Ellwood DC, Melling J, Rutter P (eds) Adhesion of microorganisms to surfaces. Academic Press, London New York San Francisco, p 57
31. Bott TR, Miller PC (1983) J Chem Technol Biotechnol 33B:177
32. Rutter P (1979) Accumulation of organisms on the teeth. In: Ellwood DC, Melling J, Rutter P (eds) Adhesion of microorganisms to surfaces. Academic Press, London New York San Francisco, p 139

11 Economic Considerations

11.1 Comparisons of Reactor Types and Efficiencies

There are several difficulties at once apparent in attempting a useful comparison of the basic reactor types discussed in the preceding chapters: one problem pertains to the characterisation of system design, the demarcation between various bioreactor types being unclear and overlap of many features occurring. The major difference, for example, between the contact and the carrier-assisted contact processes is the presence of a small quantity of inert media as support particles in the latter, a situation easily produced in the former by the introduction of nonbiodegradable suspended solids with the influent; the floc structures of both reactor configurations are practically indistinguishable. The anaerobic sludge blanket and the contact reactor have likewise many similar features, the design differences themselves being in the placement of the settling unit, i.e., internal or external to the reaction vessel. In the case of the fixed-film processes, the borderline between expanded and fluidised bed reactors is indefinite and dependent almost exclusively upon the degree of fluidisation of the carrier particles, which in turn is governed by particle density, porosity and size, and by the fluid flow rate. Other elements must necessarily enter into any qualitative comparisons made between reactor design and mode of operation: these include the strength and complexity of the waste to be treated, the influent flow rate, temperature and pH, and diurnal, seasonal or other temporal variations of these factors. Differences between reactor performances and efficiencies therefore, are not as straightforward as may be expected from design and operational variations.

11.1.1 Overloading and Intermittent Operation

In anaerobic wastewater treatment, short-term hydraulic overloading is a problem difficult to circumvent and operational failure is frequently the outcome, a situation which in many cases necessitates an extended reactor recovery period.

In reactors with long or moderate HRTs, i.e., those dependent upon retained non-attached biomass, hydraulic overload can cause biomass washout. The problem is exacerbated where no recycle facility exists, and subsequent process failure is generally the result of the loss of the slow growing methanogenic bacteria from the system. In contrast, fixed-film reactors, and in particular the expanded and fluidised bed configurations, can accommodate even severe hydraulic overloading of raw wastewater without adverse effects. The response of an anaerobic fluidised bed reactor to hydraulic overload was reported to comprise a slight decrease in effluent pH of the order of 0.2 units and an increase in effluent COD values: return to normal operational status was immediate and complete upon cessation of overload conditions, with no subsequent adverse effects observed [1]. An anaerobic filter (downflow) reactor remained stable at loading rates of 94 kg COD $m^{-3} d^{-1}$ for 24 h periods, with return to normal performance 12–48 h after overloading [2].

The UASB reactor configuration has been demonstrated to be superior to the CSTR type of system under limited hydraulic and organic overload conditions [3]. Where unattached microbial structures are important to reactor operation, however, organic overloading may alter biomass characteristics in such a way as to reduce overall process efficiency: in CSTR operations, compaction of biomass has occurred as a consequence of high organic loading [4]. Biofilm-dependent processes tend to exhibit an increased degree of stability in this respect, although the shear forces in operation in expanded and fluidised beds, combined with the effects of severe organic overload, may distort film structure and depress substrate removal efficiency. A comparative study of fixed-film and CSTR systems in the treatment of wood hydrolysate stillage indicated that the fixed-film process was superior, achieving COD reductions of up to four times those possible in the stirred tank [5]. These observations were substantiated by Stephenson and Lester [6], who reported that COD removals in a fluidised bed reactor were twice those obtained in a comparable CSTR system at equivalent COD loadings.

Biofilm processes are likewise often energy and cost effective in comparison to suspended growth systems. Expanded bed and AFB reactor configurations are in general considered more efficient than fixed bed designs for anaerobic biodegradation [7] as the latter are limited by the size of the carrier particles required to minimise clogging in the system: the large diameter filter media (2–6 cm) accumulate thick layers of biofilm, which may be susceptible to diffusional limitations [8]. Rittmann [9] attributed the superior removal efficiencies of AFB over completely mixed and filter reactors to the even distribution of biofilm throughout the system, coupled with the plug-flow properties of fluidised beds. The greatest substrate removal at high organic loading is achieved in long, narrow AFB reactors, and removal efficiencies appear superior to other systems for the same volumetric loading, although fluidisation in these reactors may need high fluid velocities to achieve expansion and mixing. High flow also increases shear losses, although these will not be the same for all reactor designs.

Generally, fixed-film reactors can withstand great loading variations and consequently intermittent operation should be well-tolerated. Packed-bed systems are reported to recover extremely efficiently after periods of non-operation [10, 11] and recovery periods of the order of 48 h are not uncommon, even after shutdown for as long as four months [12]. The UASB reactor is considered stable after periods of down-time [10] but intermittent operation can result in process failure, which necessitates complete recycle and start-up procedures [3, 13].

Typical COD loading rates associated with the reactor configurations discussed are presented in Table 9. It must be remembered, however, that in consequence of the innumerable variations in reactor size and design, composition

162

Table 9. Reactor loading rates in the mesophilic range

Anaerobic reactor type	Typical loading rates[a] $kg\,COD\,m^{-3}\,d^{-1}$
CSTR	0.25 – 3.00
Contact	0.25 – 4.00
UASB	10.00 – 30.00
CASBER	4.00 – 5.00
RBC	0.005– 0.02 $(kg\,BOD\,m^{-3}\,d^{-1})$
Anaerobic filter	1.00 – 40.00
AFFEB	1.00 – 50.00[b]
AFB	1.00 –100.00[b]

[a] Under normal operating conditions.
[b] Estimated.

of wastestreams and the difference in construction materials, operational characteristics and other factors, comparisons can only be generalised.

The great biomass accumulation possible in fixed biofilm systems, coupled with the short hydraulic and high biological solids retention times attainable, normally allows increased loading rates and thus greater productivity. The HRT is the average time of contact between the biomass and the wastewater; the solids, or biomass retention (residence) time in the system must be long enough to permit growth of the methanogens and thus prevent their washout. The minimum SRT required to maintain methanogenic bacteria in an anaerobic digestion system has been estimated as 4 days at 35 °C [4]; this corresponds to a maximum specific growth rate of 0.27 d^{-1} with a specific decay-rate coefficient of 0.02 d^{-1} [4]. The mean retention times are generally maintained above the minimum necessary for the stability of a methanogenic bacterial population and are typically 25–30 days

Table 10. Retention times and influent strengths at 35°C for anaerobic digestion systems

Reactor type	HRT (d)	SRT[a] (d)	Feedstocks (organic)
CSTR	10 –60	10–60	High SS; low to high strength
Contact	12 –15	20	Small but significant SS; low to moderate strength, but not carbohydrate rich
UASB	0.5– 7	20	Low SS; low to high strength
Anaerobic filter	0.5–12	20	Significant biodegradable SS; low to high strength
RBC	0.4– 1	30	Low SS; low to moderate strength
CASBER	0.2– 3	20	Low SS; moderate strength
AAFEB	0.2– 5	30	Low SS; low to very high strength
AFB	0.2– 5	30	Low SS; moderate to very high strength

d   Days.
SS Suspended solids.
[a] Average values.

163

in CSTRs, with a minimum of 10–12 days under mesophilic conditions. In psychrophilic operations, retention times are of the order of seven times greater [14] and in both cases, in the completely mixed system, hydraulic equals solids retention time. The HRTs and SRTs of various types of wastestream amenable to anaerobic conversion in typical digester systems are shown in Table 10. Again, retention time is related to a number of operational parameters such as organic loading rate, and SRTs in the majority of cases can be manipulated.

11.1.2 Tolerance to Toxic Shocks

The view that anaerobic systems possess an inherent disadvantage in their intolerance of the long-term and transient toxicities of many industrial wastewaters has frequently limited the introduction of anaerobic treatment processes. The low cell-yield coefficient of anaerobic microorganisms may be disadvantageous to recovery from or acclimation to wastestream toxicants in conventional digester operations, but the protracted SRTs attainable in attached-growth reactors gives these systems an intrinsic advantage over suspended biomass reactors.

The presence of toxic materials in wastestreams can inhibit or even terminate CH_4 production; the effects, however, of many inhibitors are at least partially and frequently wholly reversible, and acclimation of bacteria to particular toxicants can allow microbial exposure to inhibitory levels of around 50 times those causing severe metabolic effects in unacclimated systems [15]. Anaerobic filter reactors have been reported to be extremely stable under shock loadings of toxic substances such as petrochemicals and heavy metals [16], sulphides and disinfectants [12], formaldehyde and chloroform [11, 17] and caustic soda and strong acid [18]. Little sensitivity to daily fluctuations in influent quality indicates that packed bed systems are inherently tolerant and acclimate rapidly to changes in internal environmental conditions [19], whereas the CSTR and other suspended growth processes are frequently destabilised by even low increases in influent toxicity: a loading of 100 mg 1^{-1} formaldehyde in a CSTR system has been observed to cause reactor failure [16]. In the contact and carrier-assisted contact configurations and the UASB, damage to the floc/pellet structure of the biomass is the most frequent response to slugs of inhibitory materials; compaction of the biomass results and process efficiency and reactor stability decrease.

It was observed in one investigation that an expanded bed reactor experienced process breakdown subsequent to exposure to toxic pulp mill waste, whereas a packed bed system recovered after similar exposure [20]. Shock loadings of orthophosphoric acid, a common component in steel cleaning and descaling in the meat industry [21] were reported to be well-tolerated by AFB reactors, but CSTR systems experienced a breakdown in acidogenesis [6]. The biofilm nature of packed, expanded and fluidised beds, and in particular the protective nature of the matrix material, ensure that these anaerobic systems are extremely tolerant of both the transient and the chronic alterations in toxic content of wastewater influent. The greater biomass accumulation possible on the smaller media particles utilised in the AFB reactor probably enhances the tolerance in this type of system and high process rates of toxic wastes can be attained.

164

11.2 Wastewater Characteristics

The efficiency of anaerobic digestion operations depends to a marked extent on the characteristics of the wastestream to be treated. The diverse sources of wastewater produced include those from food processing, chemical and allied manufacturing industries, fuel production and municipal outlets. Such a wide source range must necessarily produce wastes equally diverse in composition and additionally in flow and diurnal and seasonal characteristics. The variability of wastewaters is accentuated by the distance travelled to the treatment plant: waste channels serving industrial operations are typically less than one kilometre in length, and flow and loading fluctuations mirror manufacturing processes, whereas municipal wastestreams are subject to a high degree of mixing and dilution through the sewerage system prior to arrival at the treatment plant [22].

Wastewater from the fuel production industries such as petroleum refining and coal processing can contain high levels of oil, organic and inorganic suspended solids, ammonia, sulphides, chlorides, mercaptans, cyanide, spent caustic, heavy metals, arsenic, monohydric-, polyhydric-, and multimethylated phenols, phosphates and aromatic amines. These are present in varying concentrations at various stages of refining and product finishing and are amenable to biological degradation. The inhibition of aerobic organisms utilised in biological treatment of coal wastewaters is frequently attributed to several constituents of the waste. In one investigation, thiocyanate was poorly degraded and inhibited the aerobic conversion of phenol, and cyanide, thiocyanate and phenol were found to inhibit the nitrification of ammonia [23]. Alternatively, an anaerobic environment had provided rapid acclimation to shock loadings of phenol, catechol, resorcinol, nitrobenzene and cyanide.

In industrial wastestreams, the major forms of nitrogen are ammonia and organic nitrogenous compounds such as proteins and amino acids, which yield ammonia upon biodegradation. Anaerobic bacteria can function at ammonia levels in excess of 6000 mg l^{-1} and at a pH of 8.0 [24]. Biological processes generally provide the most economical means of control of nitrogen in wastewater effluents [25] and anaerobic systems have several advantages over aerobic processes. These include the production of less sludge and the elimination of the power costs of aeration.

Of the aqueous emissions from the iron and steel industries, by-product coke plant wastewater and blast-furnace scrubber-water blowdown have been identified as the most significant contributors to toxic contamination [26]. These contaminants include suspended matter, ammonia, phenolic compounds, thiocyanate and free and complexed cyanides [27]. Trace organics such as polyaromatic hydrocarbons and heterocyclic nitrogenous compounds are also found, some of which produce long-term environmental impact. The compounds typically giving difficulty in conventional biological treatment processes include chlorinated aromatics, certain phenols (e.g. nitrophenol) and surfactants. Anaerobic biodegradation, particularly in fixed-film systems, has been successfully demonstrated in the treatment of certain of these types of recalcitrant substances [28].

Table 11. Wastes treated by anaerobic digesters

Reactor type	Type of waste	Ref.
Anaerobic filter	Leachate	Chian and DeWalle [38]
	Starch production	Mosey [39]
	Petrochemical	Lin Chou et al. [16]
	Bean blanching	van den Berg and Lentz [40]
	Alcohol stillage	Dahab and Young [41]
	Livestock	Newell [18]
	Municipal	Genung et al. [42]
	Coal gasification	Cross et al. [23]; Harper et al. [43]
	Food/Drink processing	Wheatley [44]
CSTR	Petrochemical	Lin Chou et al. [16]
	Sewage sludge	McConville and Maier [45]
	Bean blanching	van den Berg et al. [46]
	Agricultural	Lehmann and Wellinger [33]
	Citrus peel press liquor	Lane [3]
	Apple pomace	Jewell and Cummings [37]
	Synthetic meat	Stephenson and Lester [6]
UASB	Methanolic	Lettinga et al. [47, 48]
	Bean blanching	van den Berg et al. [49]
	Sugar beet	Heertjes et al. [20]
	Citrus peel press liquor	Lane [3]
	Potato	Christensen et al. [13]
	Chemical	van den Berg et al. [49]
Contact	Bean blanching/Potato/Pear peeling/ Rum stillage	van den Berg and Lentz [51]
	Bean blanching	van den Berg et al. [46]
	Synthetic	Anderson and Donnelly [4]
	Fruit and Vegetable processing	Lane [52]
AAFEB	Synthetic	Switzenbaum and Jewell [53]
	Synthetic	Morris and Jewell [54]
	Municipal	Jewell et al. [55]
	Whey	Switzenbaum and Danskin [56]
	Black liquor condensate	Norrman [20]
	Synthetic	Schraa and Jewell [57]
AFB	Dairy/Chemical/Food processing/ Soft drink/Heat treat liquor	Hickey and Owens [58]
	Whey permeate	Sutton and Li [59]
	Whey/Food/Chemical/Soft drink/Bakery/ Heat treat liquor	Jeris [60]
	Kraft bleach liquor	Salkinoja-Salonen et al. [61]
	Soy bean	Sutton et al. [62]
	Synthetic meat	Bull et al. [1]
	Synthetic dairy	Bull et al. [63]
	Coke plant	Nutt et al. [64]
	Primary wastewater	Switzenbaum et al. [65]
	Synthetic meat	Stephenson and Lester [6]
RBC	Synthetic	Tait and Friedman [66]
CASBER	Molasses/Sugar refining	Martensson and Frostell [67]

166

Table 12. Operating parameters and performances of anaerobic bioreactors treating various wastes

Reactor type	Load rate (kg COD m^{-3} d^{-1})	HRT (d)	Waste type	% COD reduction	Temp (°C)	Ref.
Anaerobic filter	0.0067	2.5	Thermally conditioned sewage sludge	65	35	Schwartz et al. [10]
	0.038	0.5	Thermally conditioned sewage sludge	50	35	Schwartz et al. [10]
	0.0172	1.0	Thermally conditioned sewage sludge	60	50	Schwartz et al. [10]
	0.0005	3.0	Alcohol stillage	84	30	Dahab and Young [41]
	0.002	1.5	Alcohol stillage	74	30	Dahab and Young [41]
	0.14	11.8	Pear peeling	64	35	van den Berg et al. [49]
CSTR	0.0028	26.7	Citrus peel press liquor	80	37	Lane [3]
	1.2	8.5	Bean blanching	70	35	van den Berg et al. [49]
	18.5	0.14	Synthetic meat	18	37	Stephenson and Lester [6]
	9.2	0.13	Synthetic meat	12	37	Stephenson and Lester [6]
UASB	0.005	4.0	Thermally conditioned sewage sludge	68	35	Schwartz et al. [10]
	0.018	1.0	Thermally conditioned sewage sludge	60	35	Schwartz et al. [10]
	0.005	4.0	Thermally conditioned sewage sludge	55	50	Schwartz et al. [10]
	0.0112	7.0	Citrus peel press liquor	95	37	Lane [3]
Contact	3.4	–	Fruit	87	36	Lane [52]
	10.0	5.5	Rum stillage	80	35	van den Berg et al. [49]
AAFEB	0.003	0.17	Sucrose	80	55	Schraa and Jewell [57]
	0.016	0.19	Sucrose	48	55	Schraa and Jewell [57]
	2.4	0.21	Glucose and Yeast extract	90	22	Jewell et al. [55]
	24.0	0.02	Glucose and Yeast extract	45	10	Jewell et al. [55]
AFB	3.5	2.06	Food processing	86	36	Jeris [60]
	24.1	0.31	Food processing	75	36	Jeris [60]
	13.4	4.9	Whole whey	84	35	Hickey and Owens [58]
	37.6	4.9	Whole whey	72	35	Hickey and Owens [58]
	15.0	3.5	Whole whey	71	24	Hickey and Owens [58]
	36.8	1.5	Whole whey	65	24	Hickey and Owens [58]
	4.8	0.53	Synthetic meat	71	37	Stephenson and Lester [6]
	9.5	0.53	Synthetic meat	76	37	Stephenson and Lester [6]
	6.0	1.67	Glucose	90	25	Bull et al. [68]
	12.0	1.67	Glucose	65	25	Bull et al. [68]
CASBER	3.9	2.4	Synthetic molasses	90	35	Martensson and Frostell [67]
	24.3	0.19	Sugar industry	89	35	Martensson and Frostell [67]
RBC	0.0195[a]	0.73	Synthetic	96	35	Tait and Friedman [66]

[a] As kg m^{-2} d^{-1}.

167

Chemical-laden wastewaters, discharged as a result of various manufacturing processes by industrial concerns, include anionic surfactants, oily emulsions, sulphates and organics. The specific composition of chemical wastes and particularly pharmaceutical wastes is often difficult to obtain as a consequence of the industrial restrictions imposed upon the availability of data.

Food processing wastes are highly variable in content and frequently seasonal in output; beverage and fruit and vegetable processing effluents include insoluble solids, wastes of processed juices, sugar, nitrogen, phosphates, bottle-washing fluids, and where wine is concerned, often sulphur dioxide. Sugar refining effluents from both cane and beet processing include wastewater from transport and condenser units, and may contain ammonia, volatile amino acids, sugar, mud solids, heat, ethanol, caustic cleaning agents, nitrates, chlorides, sulphates, phosphates and alkaline earth metals. Dairy industry wastes contain casein, protein, whole or skim milk, butterfat, buttermilk, cheese curd, whey and milkfat, and white meat and poultry processing effluents are high in fats, oils, greases, phosphorus, suspended matter, nitrogenous compounds, pathogens and blood. Overall, these wastes tend to be strongly organic and mainly soluble [22, 29–32].

The anaerobic conversion of certain wastes, especially those of the food processing industries, provide a unique opportunity to solve the dual problems of waste utilisation and rising fuel costs by the production of biogas from wastes [24, 33–36] and the potential recovery of marketable products [37]. Tables 11 and 12 enumerate various waste types which have been treated by anaerobic digestion processes, and provide some operational data.

11.3 Cost Analyses

The treatment of complex organic solid wastes such as primary and secondary wastewater sludges has long been accomplished by means of anaerobic biological conversion systems, but the protracted detention times and large volume units required precluded the use of the process for the treatment of high strength organic industrial wastewaters. In addition, anaerobic reactors were regarded as slow to start-up, expensive to introduce and less stable under fluctuating environmental conditions than the aerobic systems (trickling filters, activated sludge) adopted for these wastestreams. Recently, however, restrictions caused by rising power costs have warranted increased interest in the economics of the operation of anaerobic treatment systems, in conjunction with a number of additional factors:

1) the pumping costs of aeration can be eliminated, thus treatment can be achieved with an energy input considerably less than that required for some aerobic processes;
2) less excess sludge results from anaerobic metabolism as bacterial growth yield is low: removal costs of the sludge are therefore also low;
3) the biological and biochemical mechanisms of anaerobiosis have been investigated: increased elucidation of the mechanisms involved has enabled more efficient operational performances to be attained;

4) anaerobic digestion processes have been developed which take account of shut-down and intermittent operation, thus making the treatment of wastes with marked seasonal variations feasible;

5) where by-product recovery is practicable, a net energy gain in the form of CH_4 may be generated, especially when digester temperature can be maintained by waste heat from allied processes, or when the effluent is warm; and

6) the development of systems such as expanded and fluidised beds, with high volumetric load capacities, short HRTs and long SRTs has resulted in a reduction in reactor unit volumes and hence land area required, and has also provided the possibility of treating less-concentrated wastewaters.

Almost any waste that can be treated aerobically can be treated anaerobically [69] and the main considerations of the use of one system over the other must therefore be those of efficiency, time and cost; the outlook for the anaerobic treatment of industrial organic wastestreams appears promising. The aerobic conversion of 1 kg COD requires 2 kWh of electricity and produces 0.5 kg sludge (dry weight). Anaerobically, 1 kg COD gives rise to 0.5 m^3 biogas (i.e., 0.4 l approximately of liquid fuel) and 0.1 kg sludge (dry weight) which can be directly dewatered if required [36]. Additionally, advances in reactor technology have overcome many of the traditional drawbacks of the anaerobic treatment process. These improvements include the ability to vary the HRT and the SRT independently, allowing long SRTs to be sustained with great hydraulic flowthrough rates. The reduction of start-up time by recycle and biological controls such as inoculum cultivation, and advanced systems instrumentation have likewise increased reactor stability to environmental fluctuations. The availability of less expensive and more suitable packing and construction materials, plus carrier media, and the increase in plant operational experience have added to the practicability of the anaerobic biological system for industrial wastewater treatment.

In an economic context, many industries cannot now tolerate the increased costs of the aerobic treatment of their wastestreams. Anaerobic techniques, offering the prospect of a reduction in total costs of at least 50% [36] will more frequently be applied, especially in those industries emitting warm and concentrated effluents. The optimisation of microbial performance to enhance utilisable biogas yields at low capital cost requires careful digester design and accumulation of data on operational conditions. As more full-scale systems are implemented in industrial situations, more information will be forthcoming for the economic appraisal of anaerobic conversion processes.

The construction of a simple anaerobic digestion system de novo entails the consideration of a multiplicity of factors, not all of which are economic. The site of the reactor, the land area required and the proximity to the source of the wastestream can comprise major concerns. Land area must take into account reactor type, size and layout, production, storage and transport facilities, drainage, sludge disposal, accessory equipment and instrumentation, insulation, heating and any pre- or post-treatment necessary. The continuous and optimal use of biogas generated must be considered: the transport by pipeline or other means of useful off-gases to any great distance somewhat decreases the economic feasibility of such operations. The integration of an anaerobic process into existing waste conversion systems necessitates careful re-design of plant facilities to ensure the maximum operational efficiency for the minimum capital outlay.

The conventional unstirred tank anaerobic digesters have the advantages of simple design, straightforward construction and installation, and low cost. They are extremely efficient as conversion systems for wastestreams with high concentrations of SS. Their main drawback is the long HRT necessary for the complete degradation of complex fibrous wastes. The use of such digesters in a dual-staged thermophilic operation for the treatment of poultry wastes has proved economically feasible: for a load of 7.5 kg VS $m^{-3} d^{-1}$, at a 4 day retention time, the energy requirements in loading and temperature maintenance of the system amounted to 90 640 kcal d^{-1} (379 240 kJd^{-1}); the net output in CH_4 had an energy value estimated at 92 700 kcal d^{-1} (387 860 kJd^{-1}) i.e., a positive energy of 2 040 kcal d^{-1} (8 535 kJd^{-1}) was produced [70], the net surplus energy of the system being 50.6% of its gross value. The technique appears practicable on a commercial scale, although economic considerations are heavily influenced by biogas utilisation options.

One commercially available stirred tank reactor system has been demonstrated in the treatment of animal slurry [34]. With a handling capacity of 30 m^3 d^{-1} (8% dry solids) at 32 °C and a retention time of the order of 10 days, the biogas formed is used as produced to operate an engine-driven generator, the net electrical output of which is 33 kW. The payback period for the system has been calculated at around five years.

The generation of utilisable CH_4 from industrial and other wastes presents the opportunity of recovery of a substantial proportion of high value energy for relatively low investment and maintenance costs. The economic inadequacy of conventional treatment processes provides sufficient incentive for supplemental expansion and development of anaerobic treatment systems. The process of biomethanation, unlike other conversions, can transform most types of organic substrate. The ultimate rate of conversion activity from waste to CH_4 is a function of the number of CH_4-forming and associated bacteria: hence, the greater the retention time of the microbes in the reactor system and the greater the volume of active biomass, the higher the overall efficiency of the process will be.

Examination of a UASB reactor system indicated that the cost-effectiveness of the digester was related to the COD of the wastewater to be treated: the greater the influent COD concentration, the higher the degree of COD reduction achieved and the greater the volume of biogas produced per unit volume (capital cost) of reactor [13]. The UASB system was considered most effective for substantially soluble high strength organic wastewaters, and a combined wastewater stream from several food processing sources provided the desired nutrient balance for anaerobic treatment.

The high rate anaerobic contact and CASBER processes were developed to treat more dilute organic wastestreams. The main problem of these designs is the poor settleability of digester solids. Capital and operating costs are relatively high due to the mechanical methods of liquid/solids separation necessary, but two commercial systems based on these configurations are available [18].

The combination of high loading capacity, inherent stability and a minimum of control and associated mechanical equipment make the packed bed or anaerobic filter reactor system a viable alternative, and several economic assessments of this type of process have been made [10, 18, 42, 71, 72]. One process (Anflow), based on an upflow filter design, was estimated to require approximately 45% of the energy necessary in a comparable aerobic activated sludge system for a design flow of 190 $m^3 d^{-1}$ of moderate-strength municipal wastewater, and around 30% of the energy required by a 3 800 $m^3 d^{-1}$ activated sludge process [42, 71]. The economics of the reactor system are primarily those related to operating costs and the recovery of valuable by-products such as CH_4, but capital costs can be further reduced by employing relatively inexpensive packing materials such as plastic for the bed and utilising the increased biomass retention capacities of porous materials. Additional potential energy recovery from the anaerobic digestion of high temperature wastestreams via the application of heat pumps has also been considered [10]. Net savings of 25% in energy costs were estimated from the use of a combination of anaerobic digestion and heat pumps in a system treating thermally conditioned decant liquors.

One major benefit in a two-staged or phased anaerobic digestion system is the production of a greater quantity of CH_4, coupled with reduced reactor volume and hence reduced capital cost, in comparison with single-staged assemblies. It has been estimated that the increased solids reduction achieved in a two-staged upflow anaerobic filter configuration resulted in a reduction of 50% in sludge disposal costs over a conventional CSTR system [72].

A full-scale downflow stationary fixed-film system in operation at a sugar plant in Northern France is on line for 90 days per year and can treat 16 t COD d^{-1} [73]. The influent sugar beet waste COD concentration is $7 g l^{-1}$ and 90% COD reductions are achieved, with a concomitant biogas production which provides 3% of the mill's energy requirements. The total energy saved in comparison of an equivalent aerobic process is reported to the 600 T.O.E. (tons of oil equivalent) per 90 day running period.

The expanded and fluidised bed concepts are extensions of the anaerobic filter process and represent more efficient designs for anaerobic conversions [74]. These units maintain greater surface areas for biofilm attachment whilst reducing the dead volume of the reactors to a minimum, by decrease of media size and increase of upflow liquid velocity. The attached biomass allows the maintenance of high SRTs, and the short HRTs thus possible reduce the reactor size. Extremely high treatment efficiencies have been reported [55, 58, 60, 75]. Optimisation of these processes include considerations of media size and density of the associated pumping costs of fluidisation and recycle; the latter may be reduced by the utilisation of small-diameter, lightweight particles as support media. Biological mass and film thickness, waste specificity and operational maintenance and control of the reactor system are additional parameters that are amenable to optimisation. Fluidisation at ambient temperatures requires less pumping energy than mesophilic or thermophilic operations: water at elevated temperatures is less dense, resulting in decreased buoyancy of the media and hence increased pumping rates.

In an energy comparison of a fluidised bed system treating various industrial wastes including those of chemical and food processing origin, a positive energy balance of 2 300 kWh d^{-1} was achieved in the AFB reactor system, in comparison to a negative energy balance of -600 to -900 kWh d^{-1} produced by an activated sludge process [60]. The fluidised bed process is hence in effect a net energy producer. Capital investment payback periods of 3–5 years have been estimated for many applications, and a full-scale economic evaluation for a bottling plant indicated a return on the capital investment in less than 2 years [58]. Wastes in the AFB type of system, depending upon their organic strengths, may be treated in hours. Although no one waste system can be universal in the treatment of industrial effluents, the fluidised bed process has been demonstrated to be capable of the efficient conversion of a wide range of wastewaters, including dairy, chemical, food and heat treat liquor effluents (see Table 12).

The energy balances of a dual-phased fluidised bed configuration indicated that a 34% reduction in operational power costs accrued; this was a consequence of a 22% reduction in overall reactor volume and a decrease in hydraulic loading rate to the second reactor of the system [75]. The methanogenic bacteria comprising the biomass of such secondary stages also permit the use of smaller-diameter carrier particles in these phased assemblies, as their growth yield is low; a decreased hydraulic loading rate for fluidisation is therefore possible. Although pumping costs may preclude the use of AFB systems in the treatment of very dilute wastewaters, their efficiency in the conversion of moderate and high strength industrial effluents and their small reactor volume make them an attractive economic alternative to conventional systems. In addition, AFB reactors, in common with most other anaerobic digesters, are vertically assembled and the surface area required for their installation is therefore small. This consideration renders anaerobic systems in general and the high rate expanded and fluidised configurations in particular, eminently suitable as on-site effluent treatment processes for established industries, where existing land space is at a premium.

The factors which influence capital and investment costs of anaerobic digestion processes, in addition to those associated with retention time and performance, differ with the size, constructional material and subsequent modifications of the reactor design, and also with the unit manufacturer. Only efficiently designed reactors, effectively installed to optimise the treatment of the particular wastestream(s) involved, will provide a proper return on the investment. New applications of anaerobic conversions will present novel problems which must be dealt with within the spheres of engineering, microbiology and economics. The large-scale production of prefabricated reactor systems is the next stage in the implementation of anaerobic digestion processes for the treatment of industrial wastewaters. As the technology evolves, more design configurations of varying complexity will be produced to increase the practical biological conversion operations already in existence. Economic appraisals of these systems will provide the data necessary for a fuller understanding of the energy-conserving and cost-effective potentials of anaerobic wastewater treatment.

References

1. Bull MA, Sterritt RM, Lester JN (1983) J Chem Technol Biotechnol 33B:221
2. Kennedy KJ, van den Berg L (1982) Water Res 16:1391
3. Lane AG (1983) Environ Technol Lett 4:349

4. Anderson GK, Donnelly T (1978) Anaerobic contact digestion for treating high strength soluble wastes. In: Mattock G (ed) New processes of wastewater treatment and recovery. Ellis Horwood, Chichester, p 75
5. Good P, Moudry R, Fluvi P (1982) Biotechnol Lett 4:565
6. Stephenson T, Lester JN (in press) Biotechnol Bioeng
7. Switzenbaum MS (1983) Water Sci Technol 15:345
8. Heijnen JJ Development of a high rate fluidised bed biogas reactor. In: Proc Eur Symp Nov 1983, Noorwijkerhaout, Netherlands, p 259
9. Rittmann BE (1982) Biotechnol Bioeng 24:1341
10. Schwartz LJ, De Baere LA, Lanz RW (1982) Biotechnol Bioeng Symp 11:463
11. Young JC, Dahab MF (1983) Biotechnol Bioeng Symp 12:303
12. Messing RA (1983) Bioenergy production and pollution control with immobilized microbes. In: Tsao GT (ed) Ann reports on fermentation processes, vol 6. Academic Press, New York London, p 23
13. Christensen DR, Gerick JA, Eblen JE (1984) J Water Pollut Control Fed 56:1059
14. Winkler M (1981) Biological treatment of wastewater. Ellis Horwood, Chichester, p 211
15. Parkin GF, Speece RE (1982) J Environ Eng Div ASCE 108(EE3):515
16. Lin Chou W, Speece RE, Siddiqi RH (1979) Biotechnol Bioeng Symp 8:391
17. Dahab MF, Young JC (1982) Retention and distribution of biological solids in fixed film anaerobic filters. In: Proc 1st Int Conf on Fixed Film Biol Processes, Kings Island, Ohio, April 1982
18. Newell PJ (1981) The use of a high rate contact reactor for energy production and waste treatment from intensive livestock units. In: Vogt F (ed) Energy conservation and use of renewable energies in the bioindustries. Pergamon, Oxford, p 395
19. Kobayashi HA, Stenstrom MK, Mah RA (1983) Water Res 17:903
20. Norrman J (1983) Water Sci Technol 15:247
21. Crandell CJ, Kerrigan JE, Rohlich GA (1971) Nutrient problems in meat industry wastewater. In: Proc 26th Ind Waste Conf, Purdue Univ, Lafayette, Indiana 1971. Ann Arbor Science, Ann Arbor Michigan, p 199
22. Barnes D, Forster CF, Hrudey SE (1984) Survey in industrial wastewater treatment, vol 2: petroleum and organic chemicals industries. Pittmann, London, p 3
23. Cross WH, Chian ESK, Pohland FG, Harper S, Kharkar S, Cheng SS, Lu F (1983) Biotechnol Bioeng Symp 12:349
24. Hobson PN (1982) Production of biogas from agricultural wastes. In: Subba Rao NS (ed) Advances in agricultural microbiology. Butterworth Scientific, London, p 523
25. Barnes D, Bliss PJ (1983) Biological control of nitrogen in wastewater treatment. E & FN Spon, London, p 130
26. Melcer H, Nutt S, Marvan I, Sutton P (1984) J Water Pollut Control Fed 56:191
27. Lue-Hung C, Lordi DT, Kelada NP (1981) AIChE Symp (209)77:144
28. Salkinoja-Salonen MS, Hakulinen R, Valo R, Apajalahti J (1983) Water Sci Technol 15:309
29. Boening PH, Larsen VF (1982) Biotechnol Bioeng 24:2539
30. Duff SJB, Kennedy KJ (1982) Biotechnol Lett 4:821
31. Riera FS, Valz-Gianinet S, Gallieri D, Sineriz F (1982) Biotechnol Lett 4:127
32. Callander IJ, Barford JP (1983) Biotechnol Lett 5:755
33. Lehmann V, Wellinger A (1981) Biogas production from full-scale on-farm digesters. In: Vogt F (ed) Energy conservation and use of renewable energies in the bioindustries. Pergamon, Oxford, p 353
34. Tapp MDJ (1981) A commercial biogas producing plant. In: Vogt F (ed) Energy conservation and use of renewable energies in the bioindustries. Pergamon, Oxford, p 473
35. Wase DAJ, Gordon S (1982) Biotechnol Lett 4:436
36. Verstraete W (1983) Biomethanation of wastes: perspectives and potentials. In: Biotech 83: Proc Int Conf on the Commercial Applications and Implications of Biotechnology. Online, London, p 725
37. Jewell WJ, Cummings RJ (1984) J Food Sci 49:407
38. Chian ESK, De Walle FB (1977) Water Res 11:295
39. Mosey F (1978) Water Pollut Control 80:273
40. van den Berg L, Lentz CP (1979) Comparison between up and downflow anaerobic fixed film reactors of varying surface-to-volume ratios for the treatment of bean blanching wastes. In: Proc 34th

Ind Waste Conf, Purdue Univ, Lafayette, Indiana 1979. Ann Arbor Science, Ann Arbor Michigan, p 319

41. Dahab MF, Young JC (1982) Biotechnol Bioeng Symp 11:381
42. Genung RK, Hancher CW, Rivera AL, Harris MT (1983) Biotechnol Bioeng Symp 12:365
43. Harper SR, Cross WH, Pohland FG, Chian ESK (1984) Biotechnol Bioeng Symp 13:401
44. Wheatley A (1983) Biomethanation and by-product recovery from effluents. In: Biotech 83: Proc Int Conf on commercial applications and implications of biotechnology. Online, London, p 761
45. McConville T, Maier WJ (1979) Biotechnol Bioeng Symp 8:345
46. van den Berg L, Lentz CP, Armstrong DW (1980) Anaerobic waste treatment efficiency comparisons between fixed film reactors, contact digesters, and fully mixed continuously fed digesters. In: Proc 35th Ind Waste Conf, Purdue Univ, Lafayette, Indiana 1980. Ann Arbor Science, Ann Arbor Michigan, p 788
47. Lettinga G, van der Geest ATh, Hobma S, van der Laan J (1979) Water Res 13:725
48. Lettinga G, de Zeeuw W, Ouborg E (1981) Water Res 15:171
49. van den Berg L, Kennedy KJ, Hamoda MF (1981) Effect of type of waste on performance of anaerobic fixed film and upflow sludge bed reactors. In: Proc 36th Ind Waste Conf, Purdue Univ, Lafayette, Indiana 1981. Ann Arbor Science, Ann Arbor, Michigan, p 686
50. Heertjes PM, Kuijvenhoven LJ, van der Meer RR (1982) Biotechnol Bioeng 24:443
51. van den Berg L, Lentz CP (1978) Food processing waste treatment by anaerobic digestion. In: Proc 32nd Ind Waste Conf, Purdue Univ, Lafayette, Indiana 1977. Ann Arbor Science, Ann Arbor Michigan, p 252
52. Lane AG (1984) Environ Technol Lett 5:141
53. Switzenbaum MS, Jewell WJ (1980) J Water Pollut Control Fed 52:1953
54. Morris JW, Jewell WJ (1981) Organic particulate removal with the anaerobic attached-film expanded-bed process. In: Proc 36th Ind Waste Conf, Purdue Univ, Lafayette, Indiana 1981. Ann Arbor Science, Ann Arbor Michigan, p 621
55. Jewell WJ, Switzenbaum MS, Morris JW (1981) J Water Pollut Control Fed 53:482
56. Switzenbaum MS, Danskin SC (1981) Anaerobic expanded bed treatment of whey. In: Proc 36th Ind Waste Conf, Purdue Univ, Lafayette, Indiana 1981. Ann Arbor Science, Ann Arbor Michigan, p 414
57. Schraa G, Jewell WJ (1984) J Water Pollut Control Fed 56:226
58. Hickey RF, Owens RW (1982) Biotechnol Bioeng Symp 11:399
59. Sutton PM, Li A (1981) Anitron system and oxitron system: high rate anaerobic and aerobic biological treatment systems for industry. In: Proc 36th Ind Waste Conf, Purdue Univ, Lafayette, Indiana. Ann Arbor Science, Ann Arbor Michigan, p 665
60. Jeris JS (1983) Water Sci Technol 15:167
61. Salkinoja-Salonen MS, Nuys E-J, Sutton PM, van den Berg L, Wheatley AD (1983) Water Sci Technol 15:305
62. Sutton PM, Li A, Evans RR, Korchin S (1982) Dorr-Oliver's fixed film and suspended growth anaerobic systems for industrial wastewater treatment and energy recovery. In: Proc 37th Ind Waste Conf, Purdue Univ, Lafayette, Indiana. Ann Arbor Science, Ann Arbor Michigan, p 667
63. Bull MA, Sterritt RM, Lester JN (1983) Water Res 17:1563
64. Nutt SG, Melcer H, Pries JH (1984) J Water Pollut Control Fed 56:851
65. Switzenbaum MS, Sheehan KC, Hickey RF (1984) Environ Technol Lett 15:189
66. Tait SJ, Friedman AA (1980) J Water Pollut Control Fed 52:2257
67. Martensson L, Frostell B (1983) Water Sci Technol 15:233
68. Bull MA, Sterritt RM, Lester JN (1984) Water Res 18:1017
69. Speece RE (1983) Environ Sci Technol 17:416
70. Steinsberger SC, Shih JCH (1984) Biotechnol Bioeng 26:537
71. Genung RK, Million DL, Hancher CW, Pitt WW Jr (1979) Biotechnol Bioeng Symp 8:329
72. Ghosh S, Sajjad A, Henry MP, Bleakney RA (1984) Biotechnol Bioeng Symp 13:351
73. BS Flocor/SGN (1984) Anaerobic fermentation of effluents from a sugar mill with the SGN fixed-film process. Bridgnorth, Shropshire, UK, BS Flocor Ltd./Société Générale pour les Techniques Nouvelles
74. Switzenbaum MS (1983) Water Sci Technol 15:345
75. Sutton PM, Li A (1983) Water Sci Technol 15:333

List of Abbreviations

AAFEB	Anaerobic attached-film expanded bed
ADP	Adenosine 5'-diphosphate
AFAC	Anaerobic fluidised activated carbon
AFB	Anaerobic fluidised bed
ATP	Adenosine 5'-triphosphate
BOD	Biological oxygen demand
CASBER	Carrier-assisted sludge bed reactor
COD	Chemical oxygen demand
CSTR	Continuously stirred tank reactor
DOC	Dissolved organic carbon
GAC	Granular activated carbon
HRT	Hydraulic retention time
NAD^+	Nicotinamide adenine dinucleotide, oxidized form
$NADH_2$	Nicotinamide adenine dinucleotide, reduced form
$NAD(P)H_2$	Nicotinamide adenine dinucleotide phosphate, reduced form
OHPA	Obligate hydrogen-producing acetogenic
PAC	Powdered activated carbon
PVC	Polyvinyl chloride
RBC	Rotating biological contactor
SRT	Solids retention time
SS	Suspended solids
SVI	Sludge volume index
TOC	Total organic carbon
UASB	Upflow anaerobic sludge blanket
UVA	Unionised volatile acids
VFA	Volatile fatty acids
VS	Volatile solids
VSS	Volatile suspended solids

Subject Index

ATP (adenosine 5′-triphosphate) 13, 14, 29, 30, 54, 66
ATPase 54
Acclimation (adaptation) 60, 71, 72, 73, 74, 78, 86, 89, 103, 113, 116, 125, 135, 140, 151, 152, 156, 164, 165
Acetaldehyde 89
Acetate 1–3, 9, 11, 14, 15, 16, 19, 21, 24, 25–26, 29–36, 52, 55, 62, 63, 66, 71, 72, 79, 87, 88, 137, 144, 152, 154
Acetate kinase (acetokinase) 14, 30
Acetoacetate 13
Acetoin 31, 32
Acetoin dehydrogenase 32
Acetone 30
Acetyl CoA 12–13, 30, 77, 87, 88
Acetyl phosphate 14
Acid- 164
  base systems 63
  extraction 54
Acrolein 85
Acrylonitrile 86
Actinomyces spp. 45
Adenosine nucleotides 54
Adhesion- 40–43, 45, 65, 152, 156, 157, 159
  number 65
Adipate 87
Adsorption 53, 75, 77, 138
α-Alanine 14
β-Alanine 14
Alcohol(s) 11, 13, 26, 72
Alcohol dehydrogenase 77
Aldehydes 11
Aldrin 80, 82, 83
Alkali 71
Alkaline earth metals 71, 168
Alkalinity 63, 74, 114, 131, 136, 138, 157
Alkanols 71
Alkylbenzene sulphonates 8
Amines 165
Amino-
  acids 1, 2, 11–14, 24, 26, 27, 29, 66, 72–73, 75, 165, 168
  butyrate 14

groups 49, 53, 89
  transferase 14
Ammonia 3, 21, 22, 24, 28, 29, 63, 71, 73–74, 111, 140, 141, 165, 168
Amylase 22, 23, 26, 28
Amylopectin 28
Anaerobic filter reactor 15, 17, 43, 53, 62, 63, 66, 73, 76, 79, 86, 96, 104, 107–103, 118, 128, 135, 140, 141, 143, 150, 152, 153, 155, 156, 162, 163, 164, 166–167, 170, 171
Aniline 80, 83, 85
Antagonism 79, 86
Anthraquinone 81, 83
Antibiotics 46
Arginine- 14, 29
  deaminase 14
Aromatic molecules 80–90, 165
Arrhenius' Law 60
Arsenic 165
Azo dyes 81, 83

Bacteria-
  Acetobacter spp. 62
    woodii 67
  Acetovibrio celluliticus 26
  Bacillus spp. 26, 27, 28
    cereus 28
    licheniformis 28, 29
    stearothermophilus 28, 29
    subtilis 28, 29
  Bacteroides spp. 26, 27, 28, 45, 63, 72
    ruminicola 27
  Butyribacterium spp. 30
  Caulobacters 45
  Clostridium spp. 14, 25–27, 28, 29, 30, 67
    aceticum 26, 35
    acetobutylicum 30
    aminobutyricum 14
    barkeri 67
    bifermentans 27
    butylicum 30
    butyricum 27, 28, 30
    formicoaceticum 30
    histolyticum 27
    litusburense 27

179

Hydrofluoric acid hydrolysis 55
Hydrogen-
 bonding 41, 49, 50
 conversion to CH_4 1, 2, 15, 17, 26, 33–35,
   87, 88
 interspecies transfer 12, 15, 26, 29, 32–33
 lyase 67
 peroxide 36
 sulphide 67, 73
Hydrogenase 18, 30, 34, 67
Hydrolase 25
Hydrolysis 1, 2, 4, 8–11, 21, 23, 24, 25–29, 71,
 125, 142
Hydrophilicity 43
Hydroquinone 80, 84
Hydrosulphuric acid 63
m, o, p-Hydroxybenzoate 84
p-Hydroxycinnamate 84
Hydroxyl group 48, 49
β-Hydroxymyristate 55

Incipient flocculation 49–50
Inhibition 28, 64, 71–90, 149–150, 164
Inhibitor(s) 1, 12, 51, 62, 71–90, 130, 131, 139,
 140, 149–150, 164
Inorganic cements 45
Interfacial free energy 41, 42
Intermittent operation 161–164, 169
Ionic bonding 41, 50
Iron- 22, 30, 67–68, 72, 75, 78, 79, 100
 industry 165
 sulphide 100
Isobutyrate 11, 68
Isoheptanoate 15
Isoleucine 32
Isopropanol 30
Isovalerate 11, 26, 68

Kepone 84
Keto acids 11, 14
Ketodeoxyoctonate 55
Kinetics-
 bacterial growth 3, 4–9, 55, 139, 142, 143, 144
 methanogenesis 16–19, 55
 substrate utilisation 3–9, 55, 139, 142, 143

Lacate- 11, 13, 24, 31, 144
 dehydrogenase 31
Lactic casein whey permeate 131
Land area 129, 132, 169, 172
Lead 79
Ligands 75
Lignin 81, 89, 90, 143
Lindane 81, 82, 84
Lipase 25–29
Lipids 1, 3, 9, 22, 24, 25, 26, 55
Lipopolysaccharide 55

Lithium tracer 124
Lubricants 81
Luciferin-luciferase assay 54
Lysine 29
Lysis 50

MIC (minimum inhibitory concentration) 46
Macrocysts 52
Magnesium 17, 22
Malate 31
Malonyl-CoA 32
D-Mannuronate 54
Mass transfer 39, 51, 158
Melanins 81
Membrane potential 46, 49, 77, 89
Menaquinones 34
Mercaptans 165
2-Mercaptobenzothiazole 86
2-Mercaptoethanesulphonic acid
 (see coenzyme M)
Mercury 77–78
Metal-
 oxides 49
 sulphides 72–73
Methanol 15, 61, 62, 72, 79, 104, 151, 152, 153,
 156
Methanopterin 17, 18
Methionine 72, 75
o, m, p-Methoxybenzoate 84
Methoxychlor 84
Methyl chloride 84
2-Methylbutyrate 68
Methylcoenzyme M methyl reductase 67
Methylcoenzyme M reductase 17
Methylene chloride 84
Microbial-
 activity 54–56
 mass 54–56
Milk 168
Milkfat 168
Mixing 60, 64, 94, 95, 96, 97, 99, 101, 104, 114,
 116, 117, 124, 155, 159, 165
Molecular diffusion 45, 51
Molybdenum 66–67
Monod kinetics 4
Monohydroxybenzoate 88
Motility 48–49
Muramic acid 55
Mycoplasmas 14, 29
Myristic acid 26

NAD^+ (Nicotinamide adenine dinucleotide)
 13, 31
$NADH_2$ 13, 30, 79
$NAD(P)H_2$ 12, 18
N:P ratio 22
Nickel 22, 66–67, 75–76, 78, 79

181

184